Mechanics of Real Fluids

Mechanics of Real Fluids

M. Rahman
Halifax, Nova Scotia, Canada

WIT*PRESS* Southampton, Boston

M. Rahman
Halifax, Nova Scotia, Canada

Published by

WIT Press
Ashurst Lodge, Ashurst, Southampton, SO40 7AA, UK
Tel: 44 (0) 238 029 3223; Fax: 44 (0) 238 029 2853
E-Mail: witpress@witpress.com
http://www.witpress.com

For USA, Canada and Mexico

WIT Press
25 Bridge Street, Billerica, MA 01821, USA
Tel: 978 667 5841; Fax: 978 667 7582
E-Mail: infousa@witpress.com
http://www.witpress.com

British Library Cataloguing-in-Publication Data

A Catalogue record for this book is available
from the British Library

ISBN: 978-1-84564-502-1

Library of Congress Catalog Card Number: 2010931097

The texts of the papers in this volume were set individually by the authors or under their supervision.

Printed in Great Britain by Martins the Printers.

This book is dedicated to the loving memory of Sir James Lighthill, F.R.S. who gave the author tremendous inspiration and a love for fluid mechanics science.

A fluid mechanics memoir

Fluid mechanics is an important branch of applied mathematics. It has enormous applications in our real world problems. Since it originated about three centuries ago it is considered old, but because of the new development in new diverse directions it is modern. On the basis of the classical theories with solid foundation, researchers in this field have advanced the subject tremendously. Many excellent treaties on fluid mechanics are available in the literature. There are many books in this area that were written by many pioneers in the subject. We can name a few books which are written during the last two centuries and which are still useful as reference texts for the scientists, engineers and applied mathematicians.

The Modern Development in Fluid Dynamics, published by the Clarendon Press, Oxford in 1938, was an outstanding treaty under the editorship of Sydney Goldstein. *Laminar Boundary Layers*, the fluid motion memoir, published by the Clarendon Press in 1963, under the editorship of L. Rosenhead was another excellent book authored by some pioneers in fluid mechanics. *High-speed Flow* edited by Leslie Howarth is another excellent book.

The Hydrodynamics authored by Sir Horace Lamb is considered a masterpiece of fluid mechanics. *The Introduction to the Homogeneous Turbulence* published by Cambridge University Press in 1953 was written by G.K. Batchelor and it is considered as a very good book in the fluid mechanics field. Recently Sir James Lighthill wrote an excellent book on *Informal Introduction to Theoretical Fluid Mechanics* published by the Clarendon Press in 1979. This book describes the fundamental theoretical development of fluid flow problems in the real world and is considered an outstanding masterpiece for the young scientists and applied mathematicians. *The Dynamics of Upper Ocean* written by O.M. Phillips and published by Cambridge University Press is another outstanding book in the water related science. *Modern Fluid Dynamics* published by Van Nostrand Company, London in 1968 written by N.J. Curle and H.J. Davies is one of the best students' paperback editions. G.B. Whitham's book on *Linear and Nonlinear Waves* is a very good book on water wave problems. This fluid mechanics memoir is a compendium of works by many pioneering authors and research works of the author since he started publishing scientific papers in reputed journals. I have borrowed some physical concepts from my book *Water Waves: Relating Modern Theory to Engineering Applications* published by the Clarendon Press in 1995.

This book, I hope, will be suitable for the young scientists, graduate students, applied mathematicians and professional engineers. Theory is explained clearly and some applications are manifested in the book. The theory part is heavily borrowed from the standard textbooks but the applications part is completely new and hopefully the reader will appreciate my effort.

M. Rahman

Contents

Preface

Fluid mechanics is one of the most major areas of successful applications of mathematics. It can be considered as one of the branches of applied mathematics. The idea of writing this important book springs from Lighthill's *An Informal Introduction to Theoretical Fluid Mechanics* published by the Clarendon Press in 1986. Although fluid motion is concerned in both gas and liquids, as both of them are fluids, this book mainly deals with the motion of liquids in general, and water in particular. The theory of fluid mechanics has grown so considerably in recent years that study of the mechanics of fluids is important in many aspects of our real life.

In our real world all creatures live immersed in fluids (air or water) and their capability of motion through it is of crucial importance for their life style. As we know, systems of circulating fluid offer important means for distributing things where they are needed. As for example, the blood circulation in our body is vital. Similarly, the ocean is another great circulation system practically equally important to man. Energy stored as potential energy, chemical energy or heat energy becomes converted into kinetic energy in a water turbine, a gas turbine or a steam turbine, in each case by means of fluid flow action on rotating blades. Such flow is studied in order to improve the efficiency of turbines, which may also, in many cases, depend upon effective fluid motion for transferring heat quickly from one part to another in such an engine. Electric power generated by tides is the application of motion of water through the turbine. The design of structures intended to resist strong winds, river erosion, or violent sea motions requires an understanding of the forces exerted by winds, currents or waves upon stationary structures. So, comprehensive knowledge of flow of fluids is very important in all these cases. These are very complex problems and can be tackled only by advanced knowledge of boundary layer flow and turbulence because the fluid motion is usually propagated in a random fashion.

For complex fluid flow problem we must take recourse to the laboratory experiment or field experiment in association with the theory. Most fluid motions are much too complex and a computer is essential to find solutions, but still the problem is so complex that even if the largest and fastest of the modern computers may fail to obtain the correct result. Great progress with the effective study, and the effective computations have been made, however, it is realized that such progress required creative input on a continuing basis both from theory and experiment.

The study of fluid flow, especially the theory of water waves, has been the subject of intense scientific research since the days of Airy in 1845. As we have described above, it is of great practical importance to scientists and engineers from many disciplines in gaining insight into the complex systems of fluid motions in oceans.

Chapter 1 briefly outlines the content of the book and gives an overview of the specification of the fluid motion. Described in Chapter 2 are the basic equations of fluid motion from the view point of general fluid dynamics. Developments of Euler equations of motion for inviscid fluids have been described from a mathematical view point due to the fact that Euler equations form the backbone for the study of water wave motion. This chapter also considers two important concepts with regard to vortex kinematics and vortex dynamics. Some examples of practical interest are solved using the theory developed in this chapter. Navier-Stokes equations, only, are cited for examples as equations for viscous fluid motion. The philosophy behind the source, sink, singularities and circulation of water particles is explained accompanied by some examples. Physical interpretations of velocity potentials and stream functions are clearly explained.

Chapter 3 contains the concept of mechanics of real fluids. We describe the motion in axially symmetric 3-D bodies; pressure distribution and drag forces on a sphere are evaluated. Some exact solutions of Navier-Stokes equations are considered in this chapter. Very slow motions of fluids as manifested by Stokes and Oseen are explained with examples. Chapters 4 deals with the two-dimensional fluid motion in laminar boundary layers. Boundary layer equations for a variety of problems are discussed. The Von Misses transformations and Pohlhausen's method are discussed. Concepts of momentum and energy integrals are clearly explained. Boundary layer thicknesses such as displacement ä$_1$, momentum ä$_2$ and energy ä$_3$ are defined in integral forms. Flow in laminar wakes and jets are also considered.

Chapter 5 is devoted to the development of similarity technique and the perturbation method in fluid mechanics. Natural convection flow along a vertical plate is considered and its solution technique discussed by similarity analysis. The book concludes with Chapter 6 which is devoted to the theoretical development of turbulent flow. Some interesting theoretical examples are solved for the benefit of the graduate students who are working in the field of turbulence.

Some knowledge of vector calculus including the integral theorems such as Green's theorem, Stokes's theorem and divergence theorem is assumed on the part of the reader. Isotropic tensor calculus is used sparingly in some chapters. A familiarity with the Bessel functions, Legendre polynomials and hypergeometric functions is also expected.

Matiur Rahman, 2011
Halifax, Canada

Acknowledgments

The author is grateful to Natural Sciences and Engineering Research Council (NSERC) of Canada for its financial support. Thanks are extended to Professor Carlos Brebbia, Director of Wessex Institute of Technology, Southampton UK for his constant encouragement to write this book.

Dr Seyed Hossein Mousavizadegan deserves my special thanks for his assistance in drafting and designing some figures for this book. I also appreciate the assistance from Dr Adhi Susilo, Mr Rezaul Abid, Ms Rhonda Sutherland and Mrs Karen Conrod for helping me in drafting the figures contained in this book.

It is my great pleasure to acknowledge the kind permission granted by Professor Hubert J. Davies to reproduce Figures 5.4, 5.16, 6.5 and 7.2 from *Modern Fluid Dynamics*, Vol. 1 by N.J. Curle and H.J. Davies, published by Van Nostrand, London in 1968 into my book *Mechanics of Real Fluids*.

I am extremely thankful to Terry Edwards for sending me the formal letter of granting me permission to reproduce the figures 22, 23 and 29 from the book *An Informal Introduction to Theoretical Fluid Mechanics by James Lighthill, Clarendon Press, Oxford, 1986* into my new book, *Mechanics of Real Fluids by Matiur Rahman*. It is my great pleasure to acknowledge the IMA and its Executive Director, Dr David Youdan for granting me permission to use the illustrations from Lighthill's book. The Oxford University Press (the Clarendon Press) deserves my grateful appreciation for allowing me free permission to use some of the materials and illustrations contained in this book. Special thanks are given to Ms Shelagh Phillips, permission assistant at the Oxford University Press for taking troubles to verify the material included in this book and for her kind permission to use them in my new book.

I am also grateful to the referees for their kind reviews which help improve the contents of the book considerably, and to WIT Press, Southampton UK for their efforts in producing such a superb book.

CHAPTER 1

Introduction

Sir Isaac Newton

Sir Isaac Newton (1642–1727), mathematician and physicist, one of the foremost scientific intellects of all time was born at Woolsthorpe, near Grantham in Lincolnshire, United Kingdom, where he attended school. He entered Cambridge University in 1661, and was elected a Fellow of Trinity College in 1667 and Lucasian Professor of Mathematics in 1669. He remained at the university, lecturing in most years until 1696. During two to three years of intense mental effort he prepared Mathematical Principles of Natural Philosophy *commonly known as the* Principia, *although this was not published until 1687. For almost 300 years Newton has been regarded as the founding father of modern physical science, his achievements in experimental investigation being as innovative as those in mathematical research.*

1.1 Preliminary background

The main topics of this book relate to an introduction to theoretical fluid mechanics. A proper understanding of the flow of fluids is important to most of the physical problems in our real life and is an exciting area of study. Many scientific papers in learned journals and many relevant books have been published in which mathematical models have been used to correlate predicted and experimental data. Perfect correlation is the ultimate goal of the mathematical models, and although much has been achieved, there is great scope for future work. This chapter begins with some important concepts underlying the fluid flow phenomena. In this book, the analysis is derived under the basic principles of mechanics such as Newton's laws of motion and momentum and energy principles. It is assumed that the reader is familiar with these principles associated with the most elementary properties of fluids.

The inherent nonlinearity of any physical problem makes its behaviour very complicated. However, a particular fluid motion may be adequately represented by a linear model under certain realistic assumptions. Especially in the ocean wave problem, the assumption that the wave amplitude is small compared to the wavelength holds for most, but not all, oceanic wave phenomena. In a linear system the modes are uncoupled and can be classified and studied independently.

The effect of wind on surface waves was studied by many pioneers of theoretical fluid mechanics, including Lagrange, Airy [1], Stokes and Rayleigh [12]. They attempted to account for the elementary properties of surface waves in terms of perfect fluid theory. The problem of relating the rate of wave growth to the wind was first recognized by Kelvin. However, no progress was made on this problem until 1850, when Stevenson [14] made observations on surface waves on a number of lakes, and derived an empirical relationship between the 'greatest wave height' and the fetch. Seventy-five years later, Jeffreys [3] experimentally modelled the generation of waves by wind.

The first solutions to the problem of waves striking an internal interface were obtained by Stokes in 1847 [15]. The development of new instruments, the careful carrying out of experiments, and the more detailed data analysis have since revealed a variety of dynamical behaviours, which were previously unapparent and which offer a continuing challenge to the theoretical scientists. The greatest experimental contribution has been made by Long in his work on the problem of the excitation of internal waves caused by the flow over irregular beds.

In the nineteenth century, mathematicians and physicists were very interested in the study of disturbance generated by an obstacle in a stream of air, and of the forces between the air stream and the obstacle. The difficulty in finding the correct solution was not due to lack of a theory, but rather the existence of many theories formulated by the mathematical physicists. This subject is usually classified as external aerodynamics and it was a disturbingly mysterious subject before Ludwig Prandtl solved the mystery with his work on boundary layer theory from 1904 onwards.

In fluid mechanics, the physical effect of diffusion competes with other physical effects such as convection and this produces a more complicated behaviour. We shall give below a lucid analysis of how kinematic viscosity ν plays an important role in the study of boundary layer growth and separation. The definition of viscosity μ shows that it produces diffusion of momentum with a diffusivity (also sometimes called the kinematic viscosity)

$$\nu = \mu/\rho, \tag{1.1}$$

where ρ is the density of the fluid. We know that in the thermal process the expression $(\alpha t)^{\frac{1}{2}}$ describes the thermal boundary layer thickness. Similarly, in fluid mechanics, a vortex sheet which initially has zero thickness, and so represents a surface of *slip* or discontinuity in tangential velocity, may be expected to develop into a continuous layer whose thickness at first growth is in proportion to

$$(\nu t)^{\frac{1}{2}}, \tag{1.2}$$

where ν has the dimension (ℓ^2/t) and so the expression (1.2) has the dimension ℓ, a typical length scale. The value of the diffusivity $\nu = \mu/\rho$ at atmospheric pressure is greater for air than for water. In a frame of reference in which the body is at rest, the effect of diffusion in the boundary layer which causes the layer thickness to increase in proportion to $(\nu t)^{\frac{1}{2}}$ is combined with the effect of convection (i.e. vortexlines moving within the fluid). If U is the velocity of the body relative to the undisturbed fluid, then U also represents a typical magnitude of the velocity of the fluid at the outer edge of the boundary layer. The vortexlines in the outer part of the boundary layer are being carried along by the fluid; thus at a velocity whose magnitude is around U, and, for a body of length (ℓ), they are swept clear of the body after a time t, which is around ℓ/U. This tends to limit the growth in the boundary layer thickness to a value proportional to

$$\left(\frac{\nu\ell}{U}\right)^{\frac{1}{2}} \tag{1.3}$$

essentially because the vorticity generated at the solid surface is swept away after a time around ℓ/U, when by the expression $(\nu t)^{\frac{1}{2}}$ it has diffused such a distance. That vorticity must, of course, be replaced by a new vorticity generated at the solid surface so that the overall strength of the vortex sheet maintains the magnitude of 'slip' needed by the external irrotational flow, and this new vorticity once more diffuses a distance proportional to $(\nu\ell/U)^{\frac{1}{2}}$ before being swept away. The expression $(\nu\ell/U)^{\frac{1}{2}}$ now tells us that the ratio of ℓ to the boundary layer thickness can be very large only if the quantity

$$(U\ell/\nu)^{\frac{1}{2}} \tag{1.4}$$

is very large. This is the square of the Reynolds number

$$R = \frac{U\ell}{\nu}, \tag{1.5}$$

a non-dimensional measure of the flow speed and of the scale of the body in relation to the magnitude of viscous diffusion effects. Evidently, the square root $R^{\frac{1}{2}}$ can be very large indeed, but values

$$R > 10^4 \tag{1.6}$$

are large enough for boundary layers to remain very thin.

It was Osborne Reynolds (1842–1912) who first introduced the important concept of the Reynolds number $R = U\ell/\nu$. His experiments from 1883 onwards showed the significance of this number [13], not for the boundary layers which were to be discovered much later by Ludwig Prandtl (1875–1955), but in relation to the chaotic form of motion known as *turbulence*. Reynolds demonstrated that for a variety of particular types of motion of fluids the onset of chaotic or turbulent motion occurred when the Reynolds number exceeded a particular critical value.

Much later, experiments also showed that, although irrotational motions are immune to turbulence, any boundary layer may become turbulent at a sufficiently large value of Reynolds number. The required value depends on various factors but is typically rather nearer to 10^6 than to the limit of 10^4 noted in the inequality $R > 10^4$. The random motions of turbulence increase rates of transfer of quantities down their gradients and thus enhance the effective diffusivity above the viscous value $\nu = \mu/\rho$. Although this increases the boundary layer thickness to a value significantly above that in the expression $(\nu\ell/U)^{\frac{1}{2}}$, the boundary layer at these Reynolds numbers continues to be very thin compared with ℓ.

Because diffusion is a slow process, its direct effect upon the growth of the boundary layers on solid boundaries in flows of fluid at speeds substantial enough to satisfy the condition in the inequality $R > 10^4$ may be extremely limited, thus allowing them to remain very thin. If that happens, the flow outside such a boundary layer may be predicted quite well with the relatively simple Euler model. On the other hand, in a wide class of flows no such conclusion can be drawn because the flow *separates* from the solid surface. It is, in fact, the counterplay of the convection and diffusion of vorticity within the boundary layer which can cause it to become separated from the solid boundary in certain circumstances.

In summary, the boundary layer shows no tendency to separate where there is an accelerating external flow; where the external flow is *weakly related*, the tendency to separate, associated with a modest rate of generation of vorticity of opposite sign at the surface, is overcome by the rate of diffusion of the main boundary layer vorticity towards that surface. On the other hand, where the external flow is *strongly retarded*, the flow separates because the rate of diffusion cannot overcome the much greater rate of generation of vorticity in the sense associated with reverse flow.

Even after the appearance of Prandtl's great paper of 1904 [10] containing all the essentials of the solution to the mystery, the same type of error continued to be published. Oseen's alternative linearization of the equation of steady viscous motion made a really valuable improvement to our knowledge of very slow motions ($U\ell/\nu < 1$). However, his use of this linearization in his book *Hydrodynamik* in 1927 to discuss the behaviour of the flows in the limit of small viscosity (i.e. $U\ell/\nu \to \infty$) was hardly valuable in 1927, since it gives results in complete disagreement with the features of the flow in this limit discovered by Prandtl; in particular, the fact that the part of the flow separation from the surface is determined by the situation in the boundary layer.

In 1883, Reynolds published his celebrated account of laboratory observations on turbulent flow. Later, with the stimulus of the development of aerodynamics, Prandtl introduced the concept of a mixing length. A more fundamental approach to the dynamics of turbulence was given by Taylor in 1935 in a series of papers to the Royal Society; this is recognized as the beginning of the modern theory of turbulence. Further advancements in the subject of turbulence were made by Batchelor [2], Kolmogorov [5] and Townsend [18]. Some remarkable observations on the structure of atmospheric turbulence were made by Taylor in 1915, but it was not until 30 years later that suitable instrumentation was available to make systematic investigations.

An informal introduction to theoretical fluid mechanics can be found in a book written by Lighthill [7] and published by the Clarendon Press, Oxford. With regard to the dynamics of the upper ocean, the reader is referred to Phillips (1966). For matters concerned with waves in fluids such as sound waves, shock waves, stratified fluids and a brief description of water waves, readers are referred to Lighthill's *Waves in Fluids* [8]. Engineering applications of water waves by Rahman [11] is an important addition to the literature of fluid mechanics.

1.2 Real and ideal fluids

In fluid mechanics, most theoretical investigations begin from the concept of a perfect fluid where two contacting layers experience no tangential forces (i.e. shearing stresses) but act on each other with normal forces (i.e. pressure only). It means that no internal resistance exists in a perfect fluid. On the other hand, the inner layers of a real fluid experience tangential as well as normal stresses. These frictional tangential forces in a real fluid describe the existence of viscosity. The theory on the motion of a perfect fluid supplies many satisfactory descriptions of a real fluid. Due to the absence of tangential forces, there exists, in general, a difference in the relative tangential velocities of the perfect fluid and the solid wall wetted by the fluid. Hence, there is a slip. The existence of tangential stress and the condition of no slip near a solid wall constitute the essential differences between a perfect and a real fluid. The concepts of vorticity and circulation are founded on the basic analysis of the rotation of a fluid particle in real fluid motions and are discussed in Chapter 2.

It is important to note that certain fluids which are of great practical importance, such as water and air, have very small coefficients of viscosity. In many instances, the motion of such fluids of small viscosity agrees very well with that of a perfect

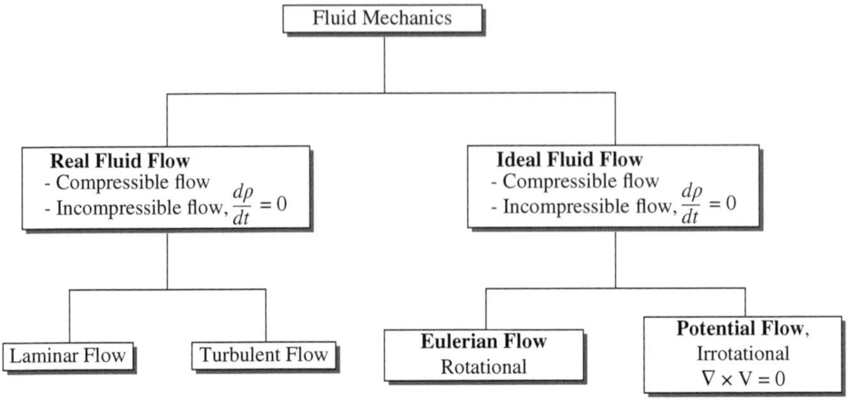

Figure 1.1: Classification of real and ideal fluid flows.

fluid, because in most cases the shearing stresses are very small. Hence, the effect of viscosity is neglected in the perfect fluid theory. A basic classification of the real and ideal fluid flows are given in Fig. 1.1.

1.3 Specification of the motion

Fluid motion is usually described in one of two ways: (a) Eulerian description of motion (i.e. *observing the fluid velocity at locations that are fixed in space*); (b) Lagrangian description of motion (i.e. *accomplished by tracking specific, identifiable fluid material volumes that are carried about with the flow*). A schematic description is shown in Fig. 1.2.

In an Eulerian description of motion, physical quantities such as the velocity $\mathbf{v}$, pressure p and density ρ are regarded as functions of the position $\mathbf{x}$ and time t. Thus $\mathbf{v} = \mathbf{v}(\mathbf{x}, t)$ and ρ represent the velocity and density of the fluid, respectively, at prescribed points in space time. In a Lagrangian description of motion, the fluid elements can be identified in terms of an initial position $\mathbf{a}$ at some initial time t_0 and the elapsed time $t - t_0$.

Thus the current position and initial position vectors are given by $\mathbf{x} = \mathbf{x}(\mathbf{a}, t - t_0)$ and $\mathbf{x_0} = \mathbf{x}(\mathbf{a})$, respectively. The velocity of a fluid element is the time derivative of its position $\mathbf{v}(\mathbf{a}, t - t_0) = \frac{\partial}{\partial t}\mathbf{x}(\mathbf{a}, t - t_0)$ so that $\mathbf{x} = \mathbf{a} + \int_{t_0}^{t} \mathbf{v}(\mathbf{a}, t - t_0)dt$. The fluid acceleration is then given by $\mathbf{f}(\mathbf{a}, t - t_0) = \frac{d\mathbf{v}}{dt} = \frac{\partial^2}{\partial t^2}\mathbf{x}(\mathbf{a}, t - t_0)$.

The total time derivative, or the derivative 'following the motion' can be expressed in Eulerian terms as (see Chapter 2) $\frac{d}{dt} = \frac{\partial}{\partial t} + (\mathbf{v} \cdot \nabla)$, the sum of the time rate of change at a fixed point and a convective rate of change.

In this book the Eulerian approach is used. Many instruments measuring fluid properties at a fixed point provide Eulerian information directly. On the other hand, in questions of diffusion or mass transport, if the motion of fluid elements is of interest, then a Lagrangian specification of the problem may be more natural. In

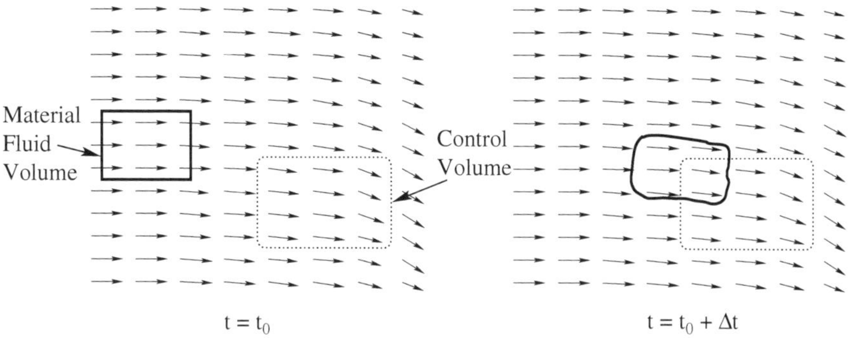

Figure 1.2: A velocity field, represented by a regular array set of velocity vectors, and within which there is a material (Lagrangian) fluid volume (solid boundary and shaded) and a control (Eulerian) volume (dotted boundary).

observation, the marking of fluid elements by dye or other traces gives Lagrangian information.

1.4 Outline of the book

This book is intended for the benefit of senior undergraduates, graduates, young scientists and engineers whose main interests are in theoretical and numerical analysis of fluid flow and its applications. The subject matter is arranged such that the topics follow in sequence, each one progressing from the previous material. Exercises at the end of each chapter are intended to give the reader experience of the principles developed in the book. The book contains plenty of solved problems.

Chapters 2–6 give the derivation of the fundamental mathematical equations. In Chapter 1, we look at the general description of the motion of fluids rather than going deep into the mathematical deduction of governing equations of motion of fluids. Chapter 2 outlines the governing equations of fluid motion to describe the physical phenomena. The coordinate systems – Cartesian, cylindrical and spherical polar – are described in developing the Navier–Stokes equations with some important classical physical problems. Chapter 3 reviews the mechanics of real fluids. This chapter portrays some classical problems with their solutions. Some solutions of Navier–Stokes equations for steady and unsteady flows are demonstrated in this chapter. Very slow motions such as Stokes's flow and Oseen's flow are considered to understand the existence of boundary layer theory.

Chapter 4 outlines the development of laminar boundary layers. Some practical problems and their analytical solutions are described in this chapter. The Von Mises transformation and the Pohlhausen's interesting series method are demonstrated with some practical problems. Some interesting problems are considered here. Chapter 5 deals with the application of similar solutions of natural convection flow with diffusion and chemical reactions, which is demonstrated clearly using the concept of boundary layer phenomenon. The book concludes with Chapter 6, which deals with turbulent flow. Reynolds equations for turbulent fluid motion and the spectral theory for homogeneous turbulence are described in this chapter.

For further information about this topic, the reader is referred to the work of Jeffreys, H. [4], Laufer, J. [6], Phillips, O.M. [9], Taylor, G.I. [16] and Thompson, S. Townsend, W. [17] listed in the reference section.

References

[1] Airy, G.B., Tides and waves. *Encyc. Metrop.*, Art. 192, pp. 241–396, 1845.

[2] Batchelor, G.K., *An Introduction to Fluid Dynamics*, Cambridge University Press: Cambridge, 1967.

[3] Jeffreys, H., On the formation of waves by winds. II. *Proc. Roy. Soc.*, **A110**, pp. 341–347, 1925.

[4] Jeffreys, H., Tidal friction in shallow seas. *Phil. Roy. Soc., London*, **A211**, pp. 239–264, 1931.

[5] Kolmogorov, A.N., The local structure of turbulence in an incompressible viscous fluid for very large Reynolds number. *C. R. Acad. Sci., USSR*, **30**, p. 301, 1941.

[6] Laufer, J., Investigation of turbulent flow in a two-dimensional channel. *Rep. Nat. Adv. Comm. Aero.*, Wash. No. 1053, 1950.

[7] Lighthill, M.J., *An Informal Introduction to Theoretical Fluid Mechanics*, Clarendon Press: Oxford, 1986.

[8] Lighthill, M.J., *Waves in Fluids*, Cambridge University Press: Cambridge, 1978.

[9] Phillips, O.M., *The Dynamics of the Upper Ocean*, Cambridge University Press: Cambridge, 1966.

[10] Prandtl, L., Uber Flussigkeitsbewegung bei sehr kleiner Reibung. Vehr. III. Tuener: Leipzig, 1904.

[11] Rahman, M., *Water Waves: Relating Modern Theory to Advanced Engineering Applications*, Clarendon Press: Oxford, 1995.

[12] Rayleigh, Lord, On the stability, or instability, of certain fluid motions. I. *Scientific Papers*, **1**, pp. 474–487, 1880.

[13] Reynolds, O., An experimental investigation of the circumstances which determine whether the motion of water shall be direct or sinuous, and of the law of resistance in parallel channels. *Phil. Trans.*, **174**, pp. 935–982, 1883.

[14] Stevenson, T., Observations on the force of waves. *Brit. Ass. (London) Rep.* See also New Edinb. Phil. J., **53(1852)**, p. 358, 1850.

[15] Stokes, G.G., On the theory of oscillatory waves. *Trans. Camb. Phil. Soc.*, **8**, pp. 441–455, 1847.

[16] Taylor, G.I., Tidal oscillations in gulfs and rectangular basins. *Proc. London Math Soc.*, **20**, pp. 148–181, 1920.

[17] Thompson, S., *Turbulence Interface Generated by an Oscillating Grid in a Stably Stratified Fluid*, Ph D dissertation, University of Cambridge, 1969.

[18] Townsend, W., *Turbulence*, Cambridge University Press: Cambridge, 1932.

CHAPTER 2

The equations of fluid motion

Sir Horace Lamb

Sir Horace Lamb (1849–1934) was born in Stockport, England in 1849, educated at Owen College, Manchester, and then Trinity College, Cambridge University. He is best known for his extremely thorough and well-written book Hydrodynamics, *which first appeared in 1876 and has been reprinted numerous times. It still serves as a compendium of useful information as well as the source of a great number of papers and books. Professor Lamb was noted for his excellent teaching and writing abilities. His research areas encompassed tides, waves, and earthquake properties as well as mathematics.*

2.1 Introduction

The study of fluid motion and other related processes, begins when the laws governing these processes have been expressed in mathematical form. Usually, in investigating the fluid flow problems, the physical situation is described by a set of differential equations, and the solution of the differential equation predicts the fluid flow pattern. For a comprehensive derivation of these equations, the reader should turn to standard textbooks including Aris [1] and Phillips (1966). In this book, we shall avoid the rigorous mathematical development of the equations.

2.2 The equations of motion

Sir Isaac Newton conceived the notion that a fluid consists of a granulated structure of discrete particles. However, the range of validity of Newton's method was limited as shown by the comparison of the theoretical and the experimental results. Later, Lagrange and Euler developed improved methods in which the fluid was regarded as a continuous medium. It is usual to adopt the Lagrangian method, where the actual paths of fluid particles are required. The Eulerian method is based on the observation of the characteristic variation of the fluid as it flows past a point previously occupied by the fluid. Thus, any quantity associated with the fluid may be functionally represented in the form $f(r, t)$.

As mentioned already, the motion of a fluid is governed by the conservation laws of mass and momentum, by the equation of state, and the laws of thermodynamics. The first of these is **the conservation of mass**,

$$\frac{d\rho}{dt} + \rho\nabla \cdot \mathbf{v} = 0 \qquad (2.1)$$

where ρ is the density of fluid, $\mathbf{v}$ the velocity vector, and t the time. Since $\frac{d}{dt} = \frac{\partial}{\partial t} + \mathbf{v} \cdot \nabla$, (2.1) can be expressed alternatively as

$$\frac{\partial\rho}{\partial t} + \nabla \cdot (\rho\mathbf{v}) = 0. \qquad (2.2)$$

If the density of fluid does not change for the element, although it may change for different element, (2.1) simplifies to

$$\nabla \cdot \mathbf{v} = 0. \qquad (2.3)$$

The momentum equation, referred to axes at rest relative to the rotating earth, takes the form

$$\rho\frac{d\mathbf{v}}{dt} + \rho\Omega \times \mathbf{v} + \nabla p - \rho\mathbf{g} = \mathbf{f}. \qquad (2.4)$$

The first term in (2.4) represents the mass-acceleration and the second the Coriolis force, in which Ω is the rotating vector, or twice the earth's angular velocity. Its magnitude

$$\Omega = |\Omega| = \frac{2\pi}{12}h^{-1} = 1.46 \times 10^{-4}s^{-1},$$

is considered constant. In the gravitational term, $\mathbf{g} = (0, 0, -g)$ represents the gravitational acceleration. The direction of $\mathbf{g}$ defines the local vertical; its magnitude varies throughout the ocean from its mean value of approximately 981 cm s^{-2} by less than 0.3%, and for dynamical purposes it can be considered constant. p is the fluid pressure. The term $\mathbf{f}$ on the right-hand side of equation (2.4) represents the resultant of all other forces acting on unit volume of the fluid. The most important of all these arises from the molecular viscosity. In almost all oceanic circumstances, where viscosity effects are important, the water can be regarded as an isotropic, incompressible Newtonian fluid, and the stress tensor

$$p_{ij} = -p\delta_{ij} + 2\mu e_{ij}, \tag{2.5}$$

where δ_{ij} is the unit tensor ($\delta_{ij} = 1$ if $i = j$, and vanishes otherwise), μ is the viscosity of the fluid and

$$e_{ij} = \frac{1}{2}\left(\frac{\partial v_i}{\partial x_j} + \frac{\partial v_j}{\partial x_i}\right) \tag{2.6}$$

is the rate of strain tensor. The frictional force per unit volume is therefore

$$f_i = 2\mu\frac{\partial e_{ij}}{\partial x_j} = \mu\left(\frac{\partial^2 v_i}{\partial x_j^2} + \frac{\partial^2 v_j}{\partial x_j \partial x_i}\right) = \mu\frac{\partial^2 v_i}{\partial x_j^2} = \mu\nabla^2\mathbf{v}, \tag{2.7}$$

from the incompressibility condition (2.3). Thus equation (2.4) will take the following familiar form

$$\rho\frac{d\mathbf{v}}{dt} + \rho\Omega \times \mathbf{v} + \nabla p - \rho\mathbf{g} = \mu\nabla^2\mathbf{v}. \tag{2.8}$$

If L is the differential length scale of a given motion in which the velocity varies in magnitude by U, the ratio $R = \frac{\rho UL}{\mu}$ (the Reynolds number) represents the relative magnitudes of the inertial and viscous terms in the momentum equation. In many fluid motions, the Reynolds number is very large, and the viscous term is often quite negligible over most of the field of motion. Specially in the oceanic problem, the viscosity term is negligible.

Two alternative forms of the momentum equation (2.4) are of interest. If the continuity equation (2.2) is multiplied by v_i and added to (2.4), there results

$$\frac{\partial}{\partial t}(\rho v_i) + \frac{\partial}{\partial x_j}(\rho v_i v_j) + \epsilon_{ijk}\rho\Omega_j v_k + \frac{\partial p}{\partial x_i} - \rho g_i = f_i \tag{2.9}$$

in the notation of Cartesian tensor. The permutation notation is defined as

$$\epsilon_{ijk} = \begin{cases} 0, & \text{if any two of } i, j, k \text{ are the same} \\ 1, & \text{if } i, j, k \text{ are an even permutation of 1, 2, 3} \\ -1, & \text{if } i, j, k \text{ are an odd permutation of 1, 2, 3} \end{cases}$$

Equation (2.9) expresses the force balance directly in terms of the rate of change of the momentum and the divergence of the momentum flux $\rho v_i v_j - p_{ij}$. Also, if $\boldsymbol{\omega} = \nabla \times \mathbf{v}$ is the vorticity of the fluid, the vector identity

$$\boldsymbol{\omega} \times \mathbf{v} = (\nabla \times \mathbf{v}) \times \mathbf{v} = \mathbf{v} \cdot \nabla \mathbf{v} - \nabla \left(\frac{1}{2} v^2 \right)$$

enables (2.4) to be expressed as

$$\rho \frac{\partial \mathbf{v}}{\partial t} + \rho(\Omega + \boldsymbol{\omega}) \times \mathbf{v} + \nabla p + \rho \nabla \left(\frac{1}{2} v^2 \right) - \rho \mathbf{g} = \mathbf{f}. \tag{2.10}$$

When ρ is in effect constant, two of the terms of this equation combine to give $\nabla(p + \frac{1}{2}\rho v^2)$, the gradient of total pressure. The Eulerian mass acceleration is therefore given as a balance between this, the gravitational force the and viscous force, and the term $\rho(\Omega + \boldsymbol{\omega}) \times \mathbf{v}$, which can be called the total vortex force, is analogous to the Coriolis force in ordinary mechanics.

If the fluid is sea water, then we know that it is a chemical solution, its density ρ is a function of three thermodynamic variables, such as the pressure p, temperature T, and salinity S, which is the mass of dissolved solids per unit mass of sea water. The functional relation among these variables is the **equation of state**:

$$\rho = \rho(p, T, S). \tag{2.11}$$

In practice it has no simple analytical form except various empirical approximations.

Laws of thermodynamics

The entropy η (a measure of disorder indicating the amount of energy that, rather than being concentrated, has become more evenly distributed and so cannot be used to do work within a particular system) per unit mass is defined so that

$$\rho \frac{d\eta}{dt} = \frac{(Q - \nabla \cdot \mathbf{h})}{T}, \tag{2.12}$$

where Q is the rate of generation of heat by friction and other irreversible processes, $\mathbf{h}$ is the heat flux and T the absolute temperature. Finally, the conservation of the dissolved solids can be expressed as

$$\rho \frac{dS}{dt} + \nabla \cdot \mathbf{s} = 0. \tag{2.13}$$

where $\mathbf{s}$ is the flux density of salt (mass per unit area per unit time).

The expressions for the molecular fluxes **h** and **s** in terms of temperature and salinity gradient in the fluid have been given by Landau and Lifshitz [7]. When the gradients are sufficiently small

$$\mathbf{h} = \beta\mathbf{s} - \kappa\nabla T,$$
$$\mathbf{s} = -\rho D\left(\nabla S + \frac{k_T}{T}\nabla T + \frac{k_p}{p}\nabla p\right), \tag{2.14}$$

where D is the diffusion coefficient and κ the thermal conductivity. The second law of thermodynamics requires that both of these coefficients be positive; the other three β, k_T, and k_p may have either sign. In pure water there is, of course, no diffusion flux, so that k_T and k_p must both vanish as $S \to 0$. In sea water at 20°C, $D \sim 1.3 \times 10^{-5}\,\mathrm{cm^2\,s^{-1}}$, and $\kappa \sim 6 \times 10^{-3}\,\mathrm{W\,cm^{-1}\,deg^{-1}}$; the values of the other coefficients are unknown.

An equation for the temperature field can be derived from these expressions. Since the temperature is a function of the state variables, ρ, η, and S, as for example,

$$\frac{dT}{dt} = \frac{\partial T}{\partial \rho}\frac{d\rho}{dt} + \frac{\partial T}{\partial \eta}\frac{d\eta}{dt} + \frac{\partial T}{\partial S}\frac{dS}{dt},$$

so that, from (2.1), (2.12) and (2.13)

$$\frac{dT}{dt} = -\rho\left(\frac{\partial T}{\partial \rho}\right)_{\eta S}\nabla\cdot\mathbf{v} + \frac{1}{\rho T}\left(\frac{\partial T}{\partial \eta}\right)_{\rho S}(Q - \nabla\cdot\mathbf{h})$$
$$-\frac{1}{\rho}\left(\frac{\partial T}{\partial S}\right)_{\eta\rho}\nabla\cdot\mathbf{s}, \tag{2.15}$$

as given by Eckart [4]. The suffixes here indicate the variables to be held constant in the partial differentiation. The last term on the right is negligible. In the first term, simple thermodynamical considerations indicate that

$$\rho\left(\frac{\partial T}{\partial \rho}\right)_{\eta S} = \frac{\gamma - 1}{\alpha} = \frac{\alpha T c^2}{C_{pS}}, \tag{2.16}$$

where α is the coefficient of thermal expansion,

$$\gamma = \frac{C_{pS}}{C_{vS}} = \frac{\text{specific heat at constant pressure}}{\text{specific heat at constant volume}}$$

the ratio of specific heats (both at constant salinity), c is the speed of sound. This term represents the cooling or heating that results from adiabatic expansion or contraction. Finally, in the second term arising from dissipation and heat flux,

$$\frac{1}{T}\left(\frac{\partial T}{\partial \eta}\right)_{\rho S} = \frac{1}{C_{vS}}, \tag{2.17}$$

so that (2.15) becomes

$$\frac{dT}{dt} = -\left(\frac{\alpha T c^2}{C_{pS}}\right)\nabla\cdot\mathbf{v} + \left(\frac{1}{\rho C_{vS}}\right)(Q - \nabla\cdot\mathbf{h}). \tag{2.18}$$

If the potential temperature θ is defined as

$$\theta = T - \int_{\rho_0}^{\rho}\left(\frac{\partial T}{\partial \rho}\right)_{\eta S} d\rho, \tag{2.19}$$

it can be seen from (2.1) and the expression above that

$$\frac{d\theta}{dt} = \frac{1}{\rho C_{vS}}(Q - \nabla\cdot\mathbf{h}); \tag{2.20}$$

in the absence of dissipation and heat transfer, the potential temperature is conserved. Since $(dP)_{\eta S} = c^2(d\rho)_{\eta S}$, (2.19) can be expressed alternatively as

$$\theta = T - \int_{p_0}^{p}\left(\frac{\partial T}{\partial p}\right)_{\eta S} dp, \tag{2.21}$$

and interpreted as the temperature of a fluid element if reduced adiabatically and with constant salinity from its ambient pressure to the reference pressure p_0.

2.3 The mechanical energy equation

The balance of mechanical energy can be found by forming an equation by scalar multiplication of the momentum equation (2.9) by the velocity vector $\mathbf{v}$, and we obtain

$$\rho\frac{d}{dt}\left(\frac{1}{2}v^2\right) + \mathbf{v}\cdot\nabla p - \rho\mathbf{v}\cdot\mathbf{g} = \mathbf{v}\cdot\mathbf{f}. \tag{2.22}$$

It is worth noting that in deducing this equation we have used the vector identity $\mathbf{v}\cdot(\Omega + \boldsymbol{\omega})\times\mathbf{v} = 0$. Now if ζ measures the vertical displacement of a fluid element (measured upward), then

$$-\rho\mathbf{v}\cdot\mathbf{g} = -\rho(u, v, w)\cdot(0, 0, -g) = \rho g w = \rho g\frac{d\zeta}{dt}$$

and with the use of the continuity equation (2.1), this equation can be expressed as

$$\frac{\partial}{\partial t}\left(\frac{1}{2}\rho v^2 + \rho g\zeta\right) + \nabla\cdot\left(\mathbf{v}(p + \frac{1}{2}\rho v^2 + \rho g\zeta)\right) - p\nabla\cdot\mathbf{v} = \mathbf{v}\cdot\mathbf{f}. \tag{2.23}$$

Here, the kinetic energy is the expression $\frac{1}{2}\rho v^2$ and the potential energy is the expression $\rho g\zeta$. The rate of change of kinetic and potential energy per unit volume is specified in terms of the divergence of the energy flux

$$\mathbf{F} = \mathbf{v}\left(p + \frac{1}{2}\rho v^2 + \rho g\zeta\right),$$

together with the rate of working in compressing the fluid and against frictional forces. If the fluid is incompressible, $\nabla \cdot \mathbf{v} = 0$; if it can be supposed inviscid, $\mathbf{f} = 0$. It is worth mentioning that the Coriolis forces do not work, since their direction is always normal to the velocity $\mathbf{v}$. They can, however, influence the energy flux indirectly.

2.4 The Boussinesq approximation

This section considers a very simplified momentum equation to derive some important formulas for the fluids. The momentum equation (2.4) reduces to a simple balance between the vertical pressure gradient and the gravitational force:

$$\frac{\partial p}{\partial z} + \rho g = 0, \tag{2.24}$$

Since in this state the fluid is isentropic, $dp = c^2 d\rho$, and (2.24) yields the density distribution as a function z

$$\rho(z) = \rho_0 \exp\left\{-g \int_0^z dz/c^2\right\}, \tag{2.25}$$

where ρ_0 is the reference density at the free surface $z = 0$ and can be taken as the mean oceanic surface density. If, in the real ocean, the variations in pressure on a fluid element are predominantly the result of variations in its depth, then $dp = -\rho g dz$. We know that $dp = c^2 d\rho$, and hence

$$\frac{d\rho}{dt} = \frac{1}{c^2}\frac{dp}{dt} = -\frac{\rho g}{c^2} w. \tag{2.26}$$

From (2.1), then

$$\nabla \cdot \mathbf{v} = \frac{g}{c^2} w. \tag{2.27}$$

In this equation, the ratio of the term on the right to the term $\frac{\partial w}{\partial z}$ on the left is of order ℓ/H where ℓ is the differential length scale of the vertical motion and $H = c^2/g$ is the local scale height for the density field. This ratio is always negligibly small, so that (2.27) can be approximated by the incompressibility condition

$$\nabla \cdot \mathbf{v} = 0. \tag{2.28}$$

In almost all applied fluid mechanics problem including oceanic motions, this is an adequate approximation to the continuity equation. The hydrostatic pressure gradient of the reference state, i.e., $\nabla p_r = \rho_r \mathbf{g}$ can be subtracted from the momentum equation (2.4), giving

$$\rho\frac{d\mathbf{v}}{dt} + \rho(\Omega \times \mathbf{v}) + \nabla \hat{p} - (\rho - \rho_r)\mathbf{g} = \mathbf{f}, \tag{2.29}$$

where $\hat{p} = p - p_r$. In the ocean water the difference $\frac{\rho - \rho_r}{\rho_r} << 1$, being of the order 10^{-3} at most. If ρ_r is replaced by ρ_0 in the inertia term, and since

$$\frac{\partial(p_r + \rho_0 g z)}{\partial z} = -g(\rho_r - \rho_0),$$

(2.29) reduces to

$$\frac{d\mathbf{v}}{dt} + \Omega \times \mathbf{v} + \frac{1}{\rho_0}\nabla p - \frac{\rho - \rho_0}{\rho_0}\mathbf{g} = \nu\nabla^2\mathbf{v}, \tag{2.30}$$

where p now represents the difference between the actual and the hydrostatic pressure in the ocean at rest with constant density ρ_0 and $\nu = \mu/\rho_0$ is the kinematic viscosity.

The set of equations (2.26, 2.28, 2.30) are called the Boussinesq approximate equations of motion. In the Boussinesq approximation, variations in the fluid density are neglected in so far as they influence the inertia; variations in the weight or buoyancy of the fluid may not be neglected.

In the Boussinesq approximate equations of motion, it is important to recognize that the gravitational acceleration and the variations in density occur only in the combination

$$\hat{b} = -g\frac{\rho - \rho_0}{\rho_0}, \tag{2.31}$$

which describes the buoyancy per unit volume. Equation (2.30) may be written in terms of buoyancy as

$$\frac{d\mathbf{v}}{dt} + \Omega \times \mathbf{v} + \frac{1}{\rho_0}\nabla p - \hat{b}\mathbf{m} = \nu\nabla^2\mathbf{v}, \tag{2.32}$$

where $\mathbf{m}$ is a unit vector vertically upwards, while from (2.26)

$$\frac{d\hat{b}}{dt} = \frac{g^2}{c^2}w. \tag{2.33}$$

If the density field be regarded as $\rho = \overline{\rho}(z) + \rho'(\mathbf{x}, t)$, where $\overline{\rho}(z)$ is the mean density and $\rho'(\mathbf{x}, t)$ is the fluctuating density about the mean, and similarly if

$$\hat{b} = B(z) + b(\mathbf{x}, t),$$

$$\begin{aligned} \text{and} \quad \frac{db}{dt} &= \left\{-\frac{\partial B}{\partial z} + \frac{g^2}{c^2}\right\} w, \\ &= \left\{\frac{g}{\rho_0}\frac{\partial\overline{\rho}}{\partial z} + \frac{g^2}{c^2}\right\} w, \\ &= -N^2 w, \quad \text{say.} \end{aligned} \tag{2.34}$$

The Brunt *Väisälä*, or stability frequency is defined as

$$N = \left\{-\frac{g}{\rho_0}\frac{\partial\overline{\rho}}{\partial z} - \frac{g^2}{c^2}\right\}^{\frac{1}{2}} = \left\{-\frac{g}{\rho_0}\frac{\partial\overline{\rho}_{pot}}{\partial z}\right\}^{\frac{1}{2}} \tag{2.35}$$

to be the natural frequency of oscillation of a vertical column of fluid given a small displacement from its equilibrium position. The fluid is statistically stable when N is real, that is, when $\frac{\partial \overline{\rho}_{pot}}{\partial z} < 0$. The distribution of N is one of the most important dynamical characteristics of the ocean. The corresponding period $\frac{2\pi}{N}$ varies usually from a few minutes in the thermocline to many hours in the deep oceans where the water is nearly neutrally stable.

An alternative form of the momentum equation (2.30), corresponding to (2.10) can be written as

$$\frac{\partial \mathbf{v}}{\partial t} + (\Omega + \boldsymbol{\omega}) \times \mathbf{v} + \nabla \left[(p/\rho_0) + \frac{1}{2} v^2 \right] - \hat{b}\mathbf{m} = \nu \nabla^2 \mathbf{v}. \tag{2.36}$$

The curl of (2.36) gives for the rate of change vorticity:

$$\frac{d\boldsymbol{\omega}}{dt} = (\Omega + \boldsymbol{\omega}) \cdot \nabla \mathbf{v} + \nabla \times (\hat{b}\mathbf{m}) + \nu \nabla^2 \boldsymbol{\omega}. \tag{2.37}$$

The terms on the right describe the generation of vorticity by stretching of the lines of total vorticity $(\Omega + \boldsymbol{\omega})$ and by the horizontal variations in buoyancy, together with the diffusion of vorticity by molecular viscosity. When the time scale of a motion is small compared with both $2\pi/\Omega(12h)$ and $2\pi/N$, the vorticity equation reduces to

$$\frac{d\boldsymbol{\omega}}{dt} = \boldsymbol{\omega} \cdot \nabla \mathbf{v} + \nu \nabla^2 \boldsymbol{\omega}. \tag{2.38}$$

the form appropriate to a homogeneous incompressible fluid in inertial frame of reference. This equation is known as Helmholtz's equation for the vorticity. The significance of the two terms on the right-hand side may be descibed as follows. Firstly, the vorticity of a given fluid element changes with time as the element is convected. Secondly, the vorticity of a fluid element is diffused by viscosity, in much the same way that heat diffuses because of conduction.

2.5 The Bernoulli equation

The corresponding momentum equation in terms of the total pressure is

$$\frac{\partial \mathbf{v}}{\partial t} + \boldsymbol{\omega} \times \mathbf{v} + \nabla \left[(p/\rho_0) + \frac{1}{2} v^2 + gz \right] = \nu \nabla^2 \mathbf{v}. \tag{2.39}$$

In many fluid motions, the influence of the viscous term in the momentum equation is quite negligible. In this situation, (2.38) shows that if, at some initial instant the vorticity of the fluid element is zero, then $\frac{d\omega}{dt} = 0$ and it remains zero. If the vorticity vanishes everywhere in the field of flow, the motion is irrotational, and, in the absence of viscous or buoyancy effects, remains so. In such flow, since $\boldsymbol{\omega} = \nabla \times v = 0$, it follows that $\mathbf{v}$ can be represented as the gradient of a scalar function, the velocity potential ϕ:

$$\mathbf{v} = \nabla \phi, \tag{2.40}$$

and, in virtue of the incompressibility condition, $\nabla \cdot \mathbf{v} = 0$, ϕ obeys Laplace's equation

$$\nabla^2 \phi = 0. \tag{2.41}$$

The term **Bernoulli's equation** is used to describe a family of first integrals of the momentum equation. In irrotational or potential flow, for instance, $\mathbf{v} = \nabla\phi$, $\boldsymbol{\omega} = 0$, $\nu = 0$ and (2.39) becomes

$$\nabla \left\{ \frac{\partial \phi}{\partial t} + \frac{p}{\rho_0} + \frac{1}{2}v^2 + gz \right\} = 0,$$

which can be integrated immediately to give

$$\frac{\partial \phi}{\partial t} + \frac{p}{\rho_0} + \frac{1}{2}v^2 + gz = f(t), \tag{2.42}$$

where $f(t)$ is an arbitrary function of time determined by the pressures imposed at the boundaries of the motion.

Another form of Bernoulli's equation can be found for steady frictionless flow, which may be stratified and rotational. It can be shown from (2.36) that in this circumstances

$$\frac{p}{\rho_0} + \frac{1}{2}v^2 + \frac{\rho}{\rho_0}gz$$

is constant along streamlines.

2.6 The Reynolds stresses

In most of the fluid mechanics, we are confronted with motions that vary in random manner in both space and time. These motions may be turbulent, or they may be associated with an irregular wave field of one kind or another, but is the randomness that is their characteristic property. The detailed velocity field and its development with time are not reproducible from experiment to experiment, though the experimental conditions are unchanged. Only the average or the statistical properties of the motion can be reproduced. The average process denoted by an over-bar is defined, it allow a separation of the motion into mean and fluctuating parts. The velocity field can be expressed as $\mathbf{V} + \mathbf{v}$, where $\overline{\mathbf{v}} = 0$. When ensemble means are taken, both $\mathbf{v}$ and $\mathbf{v}$ may be functions of x, y, z and t. The incompressibility condition is

$$\nabla \cdot (\mathbf{V} + \mathbf{v}) = 0; \tag{2.43}$$

if this equation is averaged, there results $\nabla \cdot \mathbf{v} = 0$ for the mean motion and by subtraction of this from (2.43), $\nabla \cdot \mathbf{v} = 0$ for the fluctuation velocity field also.

With the same substitution, and with the neglect of molecular viscosity, the momentum equation can be written

$$\frac{\partial}{\partial t}(\mathbf{V} + \mathbf{v}) + (\mathbf{V} + \mathbf{v}) \cdot \nabla(\mathbf{V} + \mathbf{v}) + \Omega \times (\mathbf{V} + \mathbf{v})$$
$$+ \nabla(P + p') - (B + b)\mathbf{m} = 0, \tag{2.44}$$

where the total derivative is written out fully and the pressure and the buoyancy fields are simply expressed as the sum of mean and fluctuating parts. The constant density factor ρ_0 has been incorporated into the pressure term:

$$\frac{p}{\rho_0} = P + p'.$$

The mean of equation (2.44) reduces to

$$\frac{\partial \mathbf{v}}{\partial t} + \mathbf{v} \cdot \nabla \mathbf{V} + \Omega \times \mathbf{v} + \nabla P - B\mathbf{m} = 0, \tag{2.45}$$

where, in tensor notation,

$$\begin{aligned} \Phi_i &= -u_j \overline{\frac{\partial u_i}{\partial x_j}} \\ &= - - \frac{\partial}{\partial x_j} \overline{u_i u_j} \\ &= \frac{1}{\rho_0} \frac{\partial}{\partial x_j} \tau_{ij}, \quad \text{say} \end{aligned} \tag{2.46}$$

The **Reynolds stress** $\tau_{ij} = -\rho_0 \overline{u_i u_j}$ is a second-order tensor. The influence of the turbulence on the mean flow is equivalent to that of an applied force Φ_i. So from (2.46), τ_{ij} can be interpreted as a stress set up by the fluctuating motion and acting on the mean flow, or equivalently as the mean momentum flux per unit volume carried by the velocity fluctuations.

A corresponding equation can be found to describe the mean buoyancy or density field.

$$\frac{\partial \hat{b}}{\partial t} + \mathbf{v} \cdot \nabla \hat{\mathbf{b}} = 0. \tag{2.47}$$

When the velocity field is replaced as $\mathbf{V} + \mathbf{v}$ and the buoyancy field as $B + b$ and the average taken, there results

$$\frac{\partial B}{\partial t} + \mathbf{V} \cdot \nabla B = -\nabla \cdot \mathcal{N}, \tag{2.48}$$

where

$$\mathcal{N} = \overline{\mathbf{v}b} \tag{2.49}$$

is the mean flux of buoyancy by the fluctuating motion.

2.7 Derivations of equations of motion

We have already outlined the equations of fluid motion in the previous sections. In this section, we shall briefly consider the mathematical development of some important equations, namely, the conservations of mass and momentum, also derivations of Euler's equations of motion and then Bernoulli's equation. Some special significance of irrotational motion and three-dimensional stream functions is demonstrated in this section. Navier-Stokes equations in Cartesian, cylindrical polar and spherical polar coordinates are cited using the concept of orthogonal curvilinear coordinate systems [12].

2.7.1 Conservation of mass

Let $\rho(\mathbf{r}, t)$ = density of the fluid occupying a volume V enclosed by a surface S. Then $\triangle M = \rho(\mathbf{r}, t) \triangle V$, where $\triangle M$ is the elementary mass of the fluid which has occupied an elementary volume $\triangle V$. Hence, the total mass can be obtained by integration to give

$$M = \int_V \rho(\mathbf{r}, t) dV. \tag{2.50}$$

By definition

$$\frac{dM}{dt} = \lim_{\delta t \to 0} \frac{\int_{V+\delta V} \rho(\mathbf{r}, t+\delta t) dV - \int_V \rho(\mathbf{r}, t) dV}{\delta t}.$$

However,

$$\int_{V+\delta V} \rho(\mathbf{r}, t+\delta t) dV = \int_V \rho(\mathbf{r}, t+\delta t) dV + \int_{\delta V} \rho(\mathbf{r}, t+\delta t) dV.$$

By Taylor's theorem

$$\int_V \rho(\mathbf{r}, t+\delta t) dV = \int_V \rho(\mathbf{r}, t) dV + \int_V \frac{\partial \rho}{\partial t} \delta t dV + O(\delta t^2)$$

and

$$\int_{\delta V} \rho(\mathbf{r}, t+\delta t) dV = \int_{\delta V} \rho(\mathbf{r}, t) dV + \int_{\delta V} \frac{\partial \rho}{\partial t} \delta t dV + O(\delta t^2)$$

where ρ and $\frac{\partial \rho}{\partial t}$ are functions of **r** and t only. Hence

$$\frac{dM}{dt} = \lim_{\delta t \to 0} \frac{\int_V \frac{\partial \rho}{\partial t} \delta t dV + \int_{\delta V} \rho(\mathbf{r}, t) dV + \int_{\delta V} \frac{\partial \rho}{\partial t} \delta t dV}{\delta t} + O(\delta t).$$

As $\delta t \to 0$, consequently δV must go to zero. In that situation $\int_{\delta V} \frac{\partial \rho}{\partial t} dV = 0$. Thus $\frac{dM}{dt} = \int_V \frac{\partial \rho}{\partial t} dV + \lim_{\delta t \to 0} \frac{\int_{\delta V} \rho(\mathbf{r}, t) dV}{\delta t}$. With reference to Curle and Davies [3], it may be easily shown that $\lim_{\delta t \to 0} \int_{\delta V} \frac{\rho \delta V}{\delta t} = \lim_{\delta t \to 0} \int_S \frac{\rho(lu+mv+nw)\delta t \delta S}{\delta t} = \int_S \rho \mathbf{v} \cdot \mathbf{n} dS$, where $\mathbf{v}$ is the velocity vector and $\mathbf{n}$ is the unit normal vector. Using the divergence theorem, we obtain $\lim_{\delta t \to 0} \int_{\delta V} \rho \frac{\delta V}{\delta t} = \int_V div(\rho \mathbf{v}) dV$. Therefore

$$\frac{dM}{dt} = \int_V \left[\frac{\partial \rho}{\partial t} + div(\rho \mathbf{v}) \right] dV. \tag{2.51}$$

Equation (2.51) can be deduced in the following manner also [1]. Let $\rho(\mathbf{r}, t)$ be the mass per unit volume of a homogeneous fluid at a position $\mathbf{r}$ and time t. Then the mass of any finite volume V(t) is $M = \iiint_{V(t)} \rho(\mathbf{r}, t) dV$.

We know that if the coordinate system is changed from coordinates $\mathbf{r_0}$ to the coordinates $\mathbf{r}$, the element of volume changes by the formula $dV = J \ \ dV_0$, where $J = \frac{\partial(x,y,z)}{\partial(x_0,y_0,z_0)}$ is called the Jacobian of transformations. Then we have the total derivative of M with respect to time t,

$$\begin{aligned}\frac{dM}{dt} &= \frac{d}{dt}\iiint_{V(t)} \rho(\mathbf{r}, t) dV \\ &= \frac{d}{dt}\iiint_{V_0} \rho[\mathbf{r}(\mathbf{r}_0, t), t] J dV_0.\end{aligned}$$

But we know $\frac{dJ}{dt} = (\nabla \cdot \mathbf{v})J$. Thus,

$$\begin{aligned}\frac{dM}{dt} &= \iiint_{V_0}\left[\frac{d\rho}{dt} + \rho(\nabla \cdot \mathbf{v})\right] J dV_0 \\ &= \iiint_{V(t)}\left[\frac{d\rho}{dt} + \rho(\nabla \cdot \mathbf{v})\right] dV.\end{aligned}$$

Since $\frac{d}{dt} = \frac{\partial}{\partial t} + \mathbf{v} \cdot \nabla$, we can write this formula as follows:

$$\begin{aligned}\frac{dM}{dt} &= \iiint_{V(t)}\left[\frac{\partial\rho}{\partial t} + \nabla \cdot \rho\mathbf{v}\right] dV \\ &= \iiint_{V(t)}\left[\frac{\partial\rho}{\partial t} + div(\rho\mathbf{v})\right] dV\end{aligned}$$

which is identical to (2.51).

If fluid is neither being injected nor removed from the flow field, the total mass of the fluid body must remain constant and so $\frac{dM}{dt} = 0$. Thus eqn (2.51) reduces to $\int_V [\frac{\partial\rho}{\partial t} + div(\rho\mathbf{v})] dV = 0$. But the volume V is arbitrary and consequently

$$\frac{\partial\rho}{\partial t} + div(\rho\mathbf{v}) = 0. \tag{2.52}$$

Equation (2.52) describes the basic assumption that the fluid is continuous, and is known as the **equation of continuity** for a viscous compressible flow. Equation (2.52) can be further simplified to yield

$$\frac{d\rho}{dt} + \rho div(\mathbf{v}) = 0. \tag{2.53}$$

If the fluid is incompressible, then ρ is constant and in this situation $\frac{d\rho}{dt} = 0$, and consequently eqn (2.53) reduces to

$$div \ \mathbf{v} = 0. \tag{2.54}$$

This is the equation of continuity when the fluid is incompressible.

2.7.2 Euler's equation of motion

Consider the following momentum integral

$$I = \int_V \rho \mathbf{v} dV, \tag{2.55}$$

where ρ is the density, $\mathbf{v}$ is the resultant velocity vector, δV is the elementary volume, and δS is the elementary surface area.

Here ρ dV is the elementary mass and I represents the total momentum.

The rate of change of momentum of the fluid bounded by the surface S, whether compressible or incompressible, real or ideal fluid, will be given by

$$\begin{aligned}
\frac{dI}{dt} &= \frac{d}{dt}\iiint_{V(t)} \rho \mathbf{v} dV \\
&= \frac{d}{dt}\iiint_{V_0} \rho \mathbf{v} J dV_0 \\
&= \iiint_{V_0} \left[\frac{d}{dt}(\rho \mathbf{v})J + (\rho \mathbf{v})\frac{dJ}{dt}\right] dV_0
\end{aligned}$$

However, $\frac{dJ}{dt} = (\nabla \cdot \mathbf{v})J$.

Therefore,

$$\begin{aligned}
\frac{dI}{dt} &= \iiint_{V_0} \left[\frac{d}{dt}(\rho \mathbf{v})J + (\rho \mathbf{v})(\nabla \cdot \mathbf{v})J\right] dV_0 \\
&= \iiint_{V(t)} \left[\frac{d}{dt}(\rho \mathbf{v}) + (\rho \mathbf{v})(\nabla \cdot \mathbf{v})\right] dV \\
&= \iiint_{V(t)} \left[\rho\frac{d\mathbf{v}}{dt} + \mathbf{v}\left\{\frac{d\rho}{dt} + \rho(\nabla \cdot \mathbf{v})\right\}\right] dV.
\end{aligned}$$

Since $\frac{d\rho}{dt} + \rho(\nabla \cdot \mathbf{v}) = 0$, we have

$$\frac{dI}{dt} = \iiint_{V(t)} \rho \frac{d\mathbf{v}}{dt} dV. \tag{2.56}$$

By Newton's law we know the rate of change of momentum is equal to the sum of the impressed forces. These forces are given by (a) the normal pressure thrust on the boundary S; and (b) the external force **F** per unit mass.

Therefore

$$\frac{dI}{dt} = \int_V \rho \mathbf{F} dV - \int_S p\mathbf{n} dS, \tag{2.57}$$

where **n** is the unit normal vector to δS. By using Gauss' theorem, eqn (2.57) may be written as

$$\frac{dI}{dt} = \int_V \rho \mathbf{F} dV - \int_V (\mathbf{grad}\ \ p) dV. \tag{2.58}$$

From eqns (2.56) and (2.58) we obtain

$$\int_V \left[\rho \frac{d\mathbf{v}}{dt} + \mathbf{grad}\ \ p - \rho \mathbf{F}\right] dV = 0. \tag{2.59}$$

Equation (2.59) is true for any arbitrary volume V; it follows that $\frac{d\mathbf{v}}{dt} = \mathbf{F} - \frac{1}{\rho}\mathbf{grad}\ p$, which can be written as

$$\frac{\partial \mathbf{v}}{\partial t} + (\mathbf{v} \cdot \mathbf{grad})\mathbf{v} = \mathbf{F} - \frac{1}{\rho}\mathbf{grad}\ \ p. \tag{2.60}$$

Equation (2.60) is known as **Euler's equation of motion** for an ideal compressible or incompressible fluid.

2.7.3 Bernoulli's equation revisited

If the external force **F** is conservative, that is the forces which are single valued functions of space coordinates, then **F** may be written as

$$\mathbf{F} = -\mathbf{grad}\ \ V_e, \tag{2.61}$$

where V_e is defined as the scalar potential. Also, from vector calculus

$$(\mathbf{v} \cdot \mathbf{grad})\mathbf{v} = \mathbf{grad}\left(\frac{1}{2}v^2\right) - \mathbf{v} \wedge\ \ \mathbf{curl}\ \ \mathbf{v}. \tag{2.62}$$

Assuming that the pressure is a function of density only, then Euler's equation of motion (2.60) can be written as

$$\frac{\partial \mathbf{v}}{\partial t} - \mathbf{v} \wedge\ \ \mathbf{curl}\ \ \mathbf{v} + \mathbf{grad}\left[\int \frac{dp}{\rho} + \frac{1}{2}v^2 + V_e\right] = 0. \tag{2.63}$$

Defining

$$\mathbf{grad}\left[\int \frac{dp}{\rho} + \frac{1}{2}v^2 + V_e\right] = \mathbf{grad}\ \ H, \tag{2.64}$$

and substituting from eqn (2.64) into eqn (2.63) gives

$$\frac{\partial \mathbf{v}}{\partial t} - \mathbf{v} \wedge\ \ \mathbf{curl}\ \ \mathbf{v} + \mathbf{grad}\ \ H = 0. \tag{2.65}$$

Integration of eqn (2.65) yields

$$\int \frac{dp}{\rho} + \frac{1}{2}v^2 + Ve = H, \tag{2.66}$$

where H is known as the **Bernoulli function.** Equation (2.66) is known as **Bernoulli's equation.**

Special case:

We now consider irrotational flows which are defined as flows throughout which the vorticity is zero, i.e., **curl v** $= 0$. A particular property of irrotational flows is the existence of a velocity potential which greatly facilitates their calculations. This particular property allows the Bernoulli relationship between the velocity field and the pressure field which is normally restricted to steady flows to be modified so as to become applicable in the general case of unsteady fields when these are irrotational.

When the motion is irrotational **curl v** $= 0$. For this case a scalar function $\phi(\mathbf{r}, t)$ exists such that

$$\mathbf{v} = \mathbf{grad}\ \phi, \tag{2.67}$$

where ϕ is the velocity potential. Substituting from eqn (2.67) into eqn (2.60) yields $\frac{\partial}{\partial t}(\mathbf{grad}\ \phi) + \mathbf{grad}[\int \frac{dp}{\rho} + \frac{1}{2}v^2 + V_e] = 0$. By integrating partially with respect to the space variables

$$\frac{\partial \phi}{\partial t} + \int \frac{dp}{\rho} + \frac{1}{2}v^2 + V_e = C(t), \tag{2.68}$$

where the integration constant C is a function of t.

For steady motion $\frac{\partial \phi}{\partial t} = 0$ and eqn (2.68) reduces to

$$\int \frac{dp}{\rho} + \frac{1}{2}v^2 + V_e = C, \tag{2.69}$$

where C is an absolute constant. Using $\mathbf{v} = \nabla\phi$ in the continuity eqn (2.54), it follows that the velocity potential ϕ satisfies Laplace's equation

$$\nabla^2\phi = 0. \tag{2.70}$$

2.8 The existence of irrotational motion

The vector quantity $\boldsymbol{\omega} =$ **curl v** is defined as the **vorticity vector** and if the fluid moves in such a way that the vorticity vector is zero, then the motion is said to be **irrotational**. That the irrotational motion exists in ideal fluids can be demonstrated by taking the **curl** of Euler's equation of motion (2.60) yielding $\frac{\partial \omega}{\partial t}$ - **curl**$(\mathbf{v} \wedge \boldsymbol{\omega})$ which is identically satisfied if $\boldsymbol{\omega} =$ **curl v** = 0. As a matter of fact such a motion exists for many physical problems including water wave mechanics in inviscid oceans. That the irrotational motion persists in inviscid fluids can be demonstrated below.

If C be a closed contour which moves with the fluid consisting of the same fluid particles, then the **circulation** can be defined as the integral $\Gamma = \int_C \mathbf{v} \cdot d\mathbf{r}$ where **v** is the velocity vector and $d\mathbf{r}$ the elementary length of C. This by Stokes's theorem [15] can be written as

$$\Gamma = \int\int_S \mathbf{curl}\ \mathbf{v} dS = \int\int_S \omega dS$$

where S is the surface bounded by the curve C every point of which lies within the fluid. Thus if $\boldsymbol{\omega} = 0$ is initially so will Γ be zero. From another perspective by differentiating following the motion of the fluid with respect to time we obtain

$$\frac{d\Gamma}{dt} = \int_C \frac{d\mathbf{v}}{dt} \cdot d\mathbf{r} + \int_C \mathbf{v} \cdot d\mathbf{v} = \int_C \frac{d\mathbf{v}}{dt} \cdot d\mathbf{r} + \int_C d\left(\frac{1}{2}v^2\right).$$

The second integral around the closed contour C is zero. However, using the Euler's equation (2.60) and after some simplification we can write

$$\frac{d\Gamma}{dt} = -\int_C \textbf{grad}\left(H - \frac{1}{2}v^2\right) \cdot d\mathbf{r} = -\int\int_S \textbf{curl grad}\left(H - \frac{1}{2}v^2\right) dS$$

where in the surface integral we have used the Stokes's theorem. It is now obvious that as because **curl grad** of a scalar function is identically zero, consequently we have $\frac{d\Gamma}{dt} = 0$ which expresses the fact that the circulation round a closed contour moving with the fluid is constant for all time. This result is usually known as Kelvin's theorem. In conclusion we state that the zero vorticity condition, namely, the irrotational motion exists for inviscid fluid flow motion.

2.9 Two-dimensional flow

It has been found experimentally that a large class of problems exists in which one of the velocity components, say w, is small when compared with the components u and v. Modelling such flows with the simplification obtained by setting $w = 0$ and allowing u and v to be functions of x and y, but not of z, leads to excellent agreement between theory and observation. The flow is defined as being two dimensional.

For incompressible flow in two dimensions the continuity equation, div $\mathbf{v} = 0$, where $\mathbf{v} = (u, v, 0)$, becomes

$$\frac{\partial u}{\partial x} + \frac{\partial v}{\partial y} = 0. \tag{2.71}$$

Consider the following first-order ordinary differential equation:

$$u\,dy - v\,dx = 0. \tag{2.72}$$

From the theory of first order ordinary differential equations we know that eqn (2.72) will be **exact** if the following condition is satisfied: $\frac{\partial u}{\partial x} + \frac{\partial v}{\partial y} = 0$, which is precisely the equation of continuity (2.71). Thus there exists a scalar function $\psi(x, y)$ such that

$$d\psi = u\,dy - v\,dx = 0. \tag{2.73}$$

Integrating eqn (2.73) gives

$$\psi(x, y) = \text{constant}. \tag{2.74}$$

Here $\psi(x, y)$ is a stream function, since by definition, the velocity is tangential to a streamline; therefore the differential equation of streamlines can be written as

$$\frac{dx}{u} = \frac{dy}{v}. \tag{2.75}$$

This equation can be derived from eqn (2.73). From eqn (2.73) we have $\frac{\partial \psi}{\partial x} dx + \frac{\partial \psi}{\partial y} dy = u\ dy - v\ dx$. Hence

$$u = \frac{\partial \psi}{\partial y}, \qquad v = -\frac{\partial \psi}{\partial x}. \tag{2.76}$$

It is noted that $\psi(x, y)$ is related to u and v; and also that this stream function exists only in two dimensional flow. A good account of a velocity field and its streamlines can be found in Lighthill [8].

For the case where the motion is irrotational then we must have **curl v** $= 0$, that is

$$\frac{\partial v}{\partial x} - \frac{\partial u}{\partial y} = 0. \tag{2.77}$$

Equation (2.77) can be recognized as the condition for the differential equation

$$u\ dx + v\ dy = 0, \tag{2.78}$$

to be **exact**. Thus there exists a scalar function $\phi(x, y)$ such that

$$d\phi = u\ dx + v\ dy, \tag{2.79}$$

such that $d\phi = 0$ and upon integration yields $\phi(x, y) =$ constant.

From eqn (2.79), we have $\frac{\partial \phi}{\partial x} dx + \frac{\partial \phi}{\partial y} dy = u\ dx + v\ dy$. Therefore by comparison it can be seen that

$$u = \frac{\partial \phi}{\partial x}, v = \frac{\partial \phi}{\partial y}, \tag{2.80}$$

and the velocity vector, **v** can be written as

$$\mathbf{v} = \mathbf{grad}\ \phi. \tag{2.81}$$

Here $\phi(x, y)$ is called the velocity potential. This function exists in both two- and three-dimensional flow. It is easily shown that both the stream function $\psi(x, y)$ and the velocity potential $\phi(x, y)$ satisfy Laplace's equation. Using eqn (2.76) together with the irrotational flow condition (2.77) yields

$$\frac{\partial^2 \psi}{\partial x^2} + \frac{\partial^2 \psi}{\partial y^2} = 0. \tag{2.82}$$

Similarly, using eqn (2.80) together with the continuity condition (2.71) yields

$$\frac{\partial^2 \phi}{\partial x^2} + \frac{\partial^2 \phi}{\partial y^2} = 0. \tag{2.83}$$

2.9.1 Physical interpretation of velocity potential

In practice the velocity potential ϕ is defined as the value of the line integral of the velocity vector $\mathbf{v} = (u, v, w)$ as $\phi = \int_C \mathbf{v} \cdot d\mathbf{r} = \int_C (udx + vdy + wdz)$ where C is the contour of integration. The quantity $\mathbf{v} \cdot d\mathbf{r}$ is a measure of the fluid velocity in the direction of the contour at each point. Therefore ϕ is related to the product of the velocity and length along the path between two distinct points on C. For the value of ϕ to be independent of the path, i.e., for the flow rate between these two points to be the same no matter how the integration is carried out, the term in the integrand must be an exact differential $d\phi$, so that $d\phi = udx + vdy + wdz$, and therefore, $\mathbf{v} = \mathbf{grad}\ \phi$. To ensure that this scalar function ϕ exists, it is confirmed that **curl** of the velocity vector **v** must be zero, which indeed is so. Because the vector calculus identity confirms that **curl v** = **curl grad** $\phi = 0$ always. This **curl** of the velocity vector is referred to as the vorticity $\boldsymbol{\omega}$ as described in the last section.

2.9.2 Physical interpretation of stream function

For the velocity potential, we defined ϕ in three-dimensions as the line integral of the velocity vector projected on the line element. Let us define in a similar manner the line integral composed of the velocity components perpendicular to the line element in two-dimensions as $\psi = \int_C \mathbf{v} \cdot \mathbf{n}\, dl$ where $dl = |d\mathbf{r}|$. The integrand here will physically imply that ψ represents the amount of fluid crossing the line C between two distinct points of C. The unit normal vector **n** is perpendicular to the path of integration C. This vector can be obtained from the relation that $\mathbf{n} \cdot d\mathbf{r} = 0$ such that the normal unit vector components can be obtained as $n_x = \frac{dy}{dl}$ and $n_y = -\frac{dx}{dl}$. The integral then can be written as $\psi = \int_C (udy - vdx)$. The value of this integral, i.e., the flow between these two distinct points will be independent of the path of integration provided the integrand becomes an exact differential, $d\psi$. This requires that $u = \psi_y$ and $v = -\psi_x$ from which we deduce that $u_x + v_y = 0$ which is precisely the continuity equation in two-dimensions. The ψ is defined as the stream function. It is to be noted from this mathematical analysis that there exists a stream function for two-dimensional incompressible flow. However, in general, there can be no stream function for three-dimensional flows with the exception of the axisymmetric flows whereas as we have seen already the velocity potential exists in any three-dimensional flow that is irrotational.

2.10 Complex potential

We have seen that the velocity components in two-dimensional flow can be related to $\psi(x, y)$ and $\phi(x, y)$ by the following equations:

$$u = \frac{\partial \psi}{\partial y} = \frac{\partial \phi}{\partial x}, \; v = \frac{\partial \phi}{\partial y} = -\frac{\partial \psi}{\partial x}. \tag{2.84}$$

The above equations are usually defined as the Cauchy-Riemann equations and enables the hydrodynamicists to utilize the powerful techniques of function of complex variable [9]. Using the Cauchy-Riemann conditions it can be easily verified that the lines of constant stream function ($\psi = \text{const.}$) and the lines of constant velocity potential ($\phi = \text{const.}$) are perpendicular. Also these conditions provide the necessary condition for the function

$$W = \phi + i\psi, \tag{2.85}$$

to be an analytic function of z, where $z = x + iy$.

The complex function W, whose real and imaginary parts are the velocity potential and stream function, respectively, is called the complex potential of the flow. W is an analytic function of z and, hence

$$\frac{dW}{dz} = \frac{\partial \phi}{\partial x} + i\,\frac{\partial \psi}{\partial x} = u - iv = q\, e^{-i\theta}, \tag{2.86}$$

where q is the speed of the fluid, and is given by

$$q = \left|\frac{dW}{dz}\right| = \sqrt{u^2 + v^2}, \tag{2.87}$$

and θ is the velocity direction relative to the real axis

$$\theta = tan^{-1}\frac{v}{u} = arg\left(\frac{dW}{dz}\right). \tag{2.88}$$

Also, at a stagnation point the fluid velocity is zero. Thus, if the complex potential W, describing the motion is known, the stagnation points can be obtained from the equation $\frac{dW}{dz} = 0$.

2.11 Flow along a stream tube

Sir Isaac Newton conceived the notion that a fluid consists of granulated structure of discrete particles. However, the range of validity of Newton's conjecture was limited, as shown by a comparison of the theoretical and experimental results. Later, Lagrange and Euler developed improved methods in which the fluid was regarded as a continuous medium. It is usual to adopt the Lagrangian method where the actual paths of the fluid particles are required. The Eulerian method is based on the observation of the characteristic variation of the fluid as it flows past a point previously occupied by the fluid. Thus any quantity associated with the fluid may be functionally represented in the form $f(r, t)$.

Using the Eulerian method, the state of the fluid flow along a *streamline* is defined as a line drawn in the fluid such that a tangent at each point of the line is the direction of the fluid velocity at that point. A *stream tube* is formed by drawing a set of such streamlines through all the points of a small, closed curve. More precisely, the streamlines of a steady flow are the paths along which fluid particles move. In fact, a particle on any one streamline remains always on that streamline. The streamlines associated with the velocity field

$$\mathbf{v}(\mathbf{x}) = (u(x, y, z), v(x, y, z), w(x, y, z))$$

represent the doubly infinite set of solutions of the differential equations

$$\frac{dx}{u} = \frac{dy}{v} = \frac{dz}{w},$$

where each expression represents the very short time dt during which a particle of fluid makes a change of position (dx, dy, dz). In unsteady flow, the paths of the particle of fluid (called pathlines) are completely different in shape of the streamlines at any one instant. As a particle moves along a pathline it is at each instant moving tangentially to each local streamline, but the pattern of those streamlines is changing in time. These pathlines associated with the velocity field

$$\mathbf{v}(\mathbf{x}, t) = (u(x, y, z, t), v(x, y, z, t), w(x, y, z, t))$$

can be determined by solving a system of ordinary differential equation

$$\frac{dx}{dt} = u$$

$$\frac{dy}{dt} = v$$

$$\frac{dz}{dt} = w.$$

In a steady flow, of course, a stream tube has unchanging shape because the motion of each particle of fluid on its boundary is directed along the streamline on which it is situated, and these lines are the bounding surface of the tube. Instantaneously, of course, a streamline pattern exists for an unsteady flow and this allows a stream tube to be defined such that the motion of each particle of the fluid on the surface of the tube is directed tangentially to the streamline on which it is situated and therefore tangentially to the tube composed of the streamlines. Thus the characteristics of this flow in one dimension will be fully defined once the pressure p, density ρ, velocity v, and the cross-sectional area A of the tube are known as functions of the axial distance of the tube. Hence, four equations are needed to evaluate these four unknowns. It should be noted here that when the cross-section A is infinitesimally small, the stream tube is known as the *stream filament*.

2.12 Vortex kinematics

We have already introduced the vorticity vector $\boldsymbol{\omega}$ in two- and three-dimensional cases. In this section, we shall try to explore a little bit more about this important entity of fluid mechanics. To get full qualitative information regarding three-dimensional motions specified in general by four unknown quantities such as the density ρ, the velocity components u, v, and w. It was, accordingly, only after another brilliant simplification has been introduced into the theoretical fluid mechanics, and developed through the work of Joseph Lagrange (1736–1831),

Augustin Cauchy (1789–1857), Hermann von Helmholtz (1821–1894), George Stokes (1819–1903), and Lord Kelvin (1824–1907), that significant progress could be made in the analysis of the forbiddingly nonlinear equations of motions described in the previous sections. As a result of their work, we recognize that the study of fluid motions lies in the related concepts of vorticity and circulation. These concepts are formed on an acceptance of the need to analyse the rotation of a particle of fluid. In the rigid body vortex motion, there are two factors, rotation and the associated angular momentum. But in the fluid motion, the fluid particle, however, exhibits a major difference from a rigid body: it does not simply rotate (in its motion relative to its mass centre) but is also deformed at the same time in certain ways.

2.12.1 Vortexlines and vortextubes

We have already defined the streamlines that are composed of joining velocity vectors. In the same way we can form the vortexlines by joining the vorticity vectors and then form vortextubes. The vorticity vector is given as

$$\nabla \times \mathbf{v} = \boldsymbol{\omega} = (\xi, \eta, \zeta) \tag{2.89}$$

such that the components of the vorticity $\boldsymbol{\omega}$ can be explicitly written as

$$\begin{aligned} \xi &= \frac{\partial w}{\partial y} - \frac{\partial v}{\partial z} \\ \eta &= \frac{\partial u}{\partial z} - \frac{\partial w}{\partial x} \\ \zeta &= \frac{\partial v}{\partial x} - \frac{\partial u}{\partial y} \end{aligned} \tag{2.90}$$

First of all we note that the vorticity field is solenoidal, which means that

$$\nabla \cdot \boldsymbol{\omega} = 0.$$

Indeed, for any vector field $\mathbf{v}$ we have $\nabla \cdot \nabla \times \mathbf{v} = div\ curl\ \mathbf{v} = 0$ which follows from the vector identity. This makes the result seem very natural as the vanishing of a scalar triple product in which the gradient operator ∇ appears twice.

To construct the vortexlines and vortextubes we follow Lighthill's work [8]. As suggested by Lighthill, we can draw large numbers of very short arrows, each proportional to the local value of the vorticity field, and join these to form an assemblage of curves called vortexlines. Mathematically, they are specified at each time t by the equations (Fig. 2.1)

$$\frac{dx}{\xi} = \frac{dy}{\eta} = \frac{dz}{\zeta} \tag{2.91}$$

Next we draw a general vortextube as shown in Fig. 2.2 composed of the vortexlines passing through a particular loop, and a very thin vortextube. For either

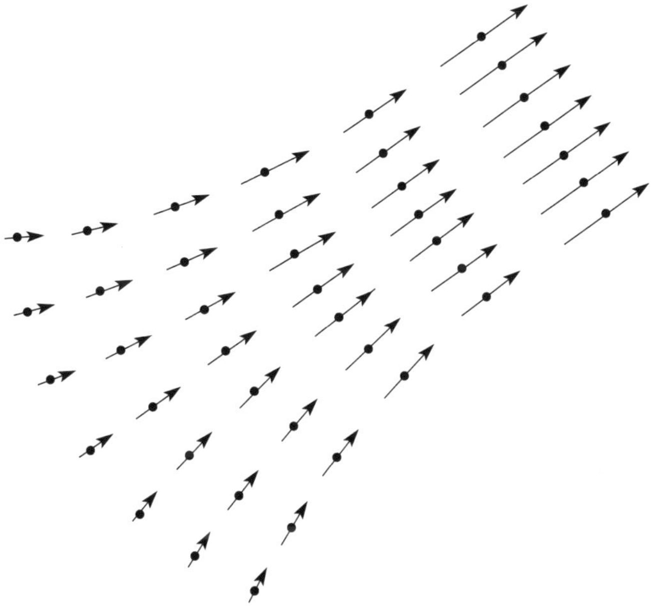

Figure 2.1: Vorticity vectors joined to form vortexlines. (From Lighthill [8], with permission from IMA.)

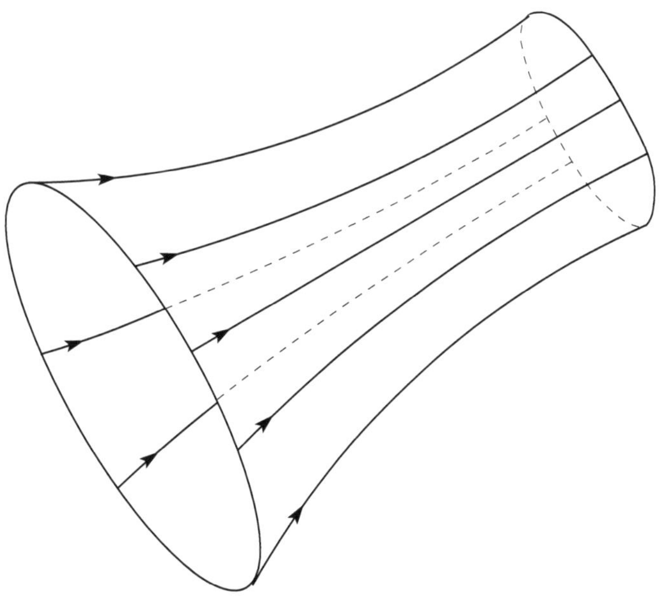

Figure 2.2: A vortextube is a surface comprising all the vortexlines passing through points of a particular loop. (From Lighthill [8], with permission from IMA.)

vortexlines or vortextube, the solenoidal property $div\ \boldsymbol{\omega} = 0$ implies that all along the vortextube a certain quantity takes an unchanging value. For the thin vortextube that quantity is expected to be the product of its cross sectional area with the magnitude of the vorticity [8].

2.12.2 Circulation

Following Lighthill we can show that the surface integral of a vortextube (Fig. 2.2) can be expressed as

$$\int\int_S \boldsymbol{\omega} \cdot \mathbf{n} dS = \int\int_S \mathbf{curl\ v} \cdot \mathbf{n} dS = \text{constant} \tag{2.92}$$

for all surfaces S spanning a given vortextube. The quantity stated in eqn (2.92) to take everywhere a constant value along a general vortextube is a characteristic property of that vortextube, sometimes called its "strength". We consider now whether any special physical interpretation can be assigned to this quantity. We know that the corresponding quantity for a general streamtube would be the integral

$$\int\int_S \mathbf{v} \cdot \mathbf{n} dS \tag{2.93}$$

across a surface spanning the streamtube. The physical interpretation of this quantity is readily seen to be the rate of volume flow along the streamtube.

It is a famous mathematical theorem due to Stokes which gives a special interpretation of the left-hand side of eqn (2.92). Stokes's theorem applies to any surface S for which we are able to make a consistent, continuously varying selection of a direction for the unit normal vector **n** at each point of the surface and for which the boundary C consists of one or more closed curves. Stokes's theorem states that, for any vector field **v**,

$$\int\int_S (\mathbf{curl\ v}) \cdot \mathbf{n} dS = \int_C \mathbf{v} \cdot d\mathbf{x} \tag{2.94}$$

Here, the integral of

$$\mathbf{v} \cdot d\mathbf{x} = u dx + v dy + w dz \tag{2.95}$$

along each of the closed curves making up a boundary C must be taken in positive sense relative to the direction chosen for **n**.

For any closed curve C, the integral of

$$\oint_C \mathbf{v} \cdot d\mathbf{x} \tag{2.96}$$

taken around C in a particular sense, given the special name **Circulation**. More precisely, it is called the circulation in that sense around C. Here the meaning of the small vector displacement $d\mathbf{x}$ along C is a small element ds of arc length along

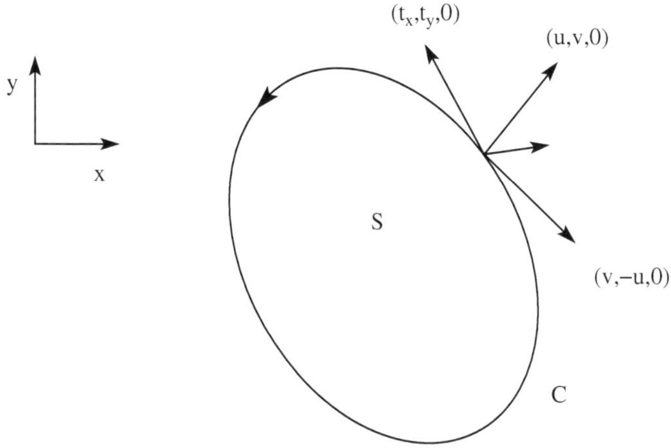

Figure 2.3: Stokes's theorem (Green's theorem) is applied here to a flat surface S with the normals to S pointing out of the paper towards the reader. A component $ut_x + vt_y$ of $(u, v, 0)$ along the tangent to C is identical with a component of the perpendicular vector $(v, -u, 0)$ along the outward normal to C. (From Lighthill [8], with permission from IMA.)

C, while its direction is that of $\mathbf{t}$, a unit tangent vector in the chosen sense. Thus the circulation (expression (2.96)) can be written

$$\oint_C (\mathbf{v} \cdot \mathbf{t})\mathbf{ds}; \tag{2.97}$$

and, it takes a positive value if, on an average with respect to arc length along the curve, positive values of the tangential resultant $\mathbf{v} \cdot \mathbf{t}$ of the velocity field out weigh negative values.

There are some dynamical reasons why circulation is important, but confine our attention here to its importance for the kinematics of vortextube (Fig. 2.2), its boundary is a single closed curve C embracing the vortextube once, and any such curve C is the boundary of a spanning surface S. Stokes's theorem tells us, then, that

$$\int\int_S \boldsymbol{\omega} \cdot \mathbf{n} dS = \oint_C \mathbf{v} \cdot d\mathbf{x}; \tag{2.98}$$

i.e. the 'strength' of the vortextube (the quantity which takes the same constant value (eqn (2.92) for any surface spanning it) is equal to the circulation around any closed curve C, which embraces it once, taken in the positive sense with respect to the direction of the normal $\mathbf{n}$.

In two-dimensional case (Fig. 2.3), to facilitate the proof of equation (2.98) for a flat surface S we are free to adopt coordinates x, y and z for which S is a part of the plane $z = 0$ and **n** is a normal unit vector in the z direction. The eqn (2.98) becomes

$$\int\int_S (v_x - u_y)dS = \oint_C (udx + vdy)$$
$$= \oint_C (ut_x + vt_y)ds \tag{2.99}$$

in terms of the z component, and the components $(t_x, t_y, 0)$ of the unit tangent vector which in the form of expression $\oint(\mathbf{v}\cdot\mathbf{t})\mathbf{ds}$. Equation (2.99) is also usually known as Green's theorem.

2.13 Vortex dynamics

The meanings and interrelationships of vorticity and circulation have been illustrated in the previous section from the pure standpoint of kinematics, which deals with the analysis of motions without considering the dynamical effect of forces. The present section explores the study of their very fruitful contribution to an understanding of the dynamics of fluid motions.

A single discovery by Kelvin in 1869 about the persistence of circulation is used to derive all the theoretical results given in this section, including some discoveries by Lagrange, Cauchy and Helmholtz. The section ends with a first attempt at giving insight into the properties of boundary layers, based on combination of theoretical understanding and experimental data.

2.13.1 The persistence of circulation

The persistence ascribed to circulation by Kelvin's theorem does not refer to the circulation around a closed curve C whose shape and position are fixed in space. The theorem is concerned, rather, with a closed curve C consisting always of the same particles of fluid. We can think of C, then, as a *necklace of particles of fluids* one whose shape and position change continually as those particles of fluids move. This is why it is usually described as a closed curve moving with the fluid.

Kelvin's theorem is exact for the Euler model, which is being derived from the momentum equation

$$\rho\frac{D\mathbf{v}}{Dt} = -grad(p + \rho gH - p_0), \tag{2.100}$$

where the material derivative is given by $\frac{D\mathbf{v}}{Dt} = \frac{\partial\mathbf{v}}{\partial t} + \mathbf{v}\cdot\nabla\mathbf{v}$. It states that the circulation around a closed curve C moving with the fluid remains constant.

Consider now a curve C which, because it is made up of a necklace of identified particles of fluid, can be described as moving with the fluid. Assuming the momentum equation (2.100) at each point of C, we deduce Kelvin's result that the rate of change of circulation around a closed curve C is zero.

If we denote circulation by the symbol Γ such that $\Gamma = \oint_C \mathbf{v} \cdot d\mathbf{x}$ then the rate of change of circulation is

$$\begin{aligned}
\frac{D\Gamma}{Dt} &= \frac{D}{Dt}\left(\oint_C \mathbf{v} \cdot d\mathbf{x}\right) \\
&= \oint_C \frac{D}{Dt}(\mathbf{v} \cdot d\mathbf{x}) \\
&= \oint_C \mathbf{v} \cdot d\mathbf{v} + \frac{D\mathbf{v}}{Dt} \cdot d\mathbf{x} \\
&= \oint_C d\left(\frac{1}{2}\right)(\mathbf{v} \cdot \mathbf{v}) + \left(-\frac{1}{\rho}\right)\nabla(p + \rho g H - p_0) \cdot d\mathbf{x} \\
&= \oint_C d\left(\frac{1}{2}|\mathbf{v}|^2\right) - \frac{1}{\rho}dp \\
&= 0.
\end{aligned} \tag{2.101}$$

This mathematical deduction simply describes the persistence of circulation around a closed curve moving with the fluid.

One especially valuable deduction from Kelvin's theorem is concerned with the movement of vortexlines. This is Helmholtz theorem which states that vortexlines move with the fluid. Such a statement has to be understood in the same sense as the statements about a closed curve C moving with the fluid as demonstrated above. It means, in fact, that the particle of fluid of which any vortexline is composed at any one instant move in such a way that the same chain of particles of fluid continue to form a vortexline at all later instant.

2.13.2 Line vortices and vortex sheets

Lighthill [8] has given a lucid description with illustrations about the line vortex and vortex sheets in his book on *An information introduction to theoretical fluid mechanics*. In this section we shall just define these two important entities of fluid mechanics without going into detailed calculations. The true condition is that a solid boundary equates not only the normal components of the velocities of the fluid and of the solid surface but also their tangential components. The effect of this is to destroy any exact permanence of irrotational flows, essentially through the generation of vorticity at any such solid surface.

Often, the convection and diffusion of vorticities, which initially emerged from the solid boundary, are found to leave and then concentrated in a very thin region. It may, in many cases, be thin in just one of its dimensions, as a sheet is, and then it is often called a *vortex sheet*. A boundary layer is an example of a vortex sheet. Alternatively, some vortexlines that have been convected away from the solid boundary may, in certain circumstances, become concentrated in a region thin in two of its dimensions (essentially in the immediate neighbourhood of a single line), and such a concentration of vorticity around a single line is usually called a *line vortex* . A schematic diagram is depicted in Fig. 2.4.

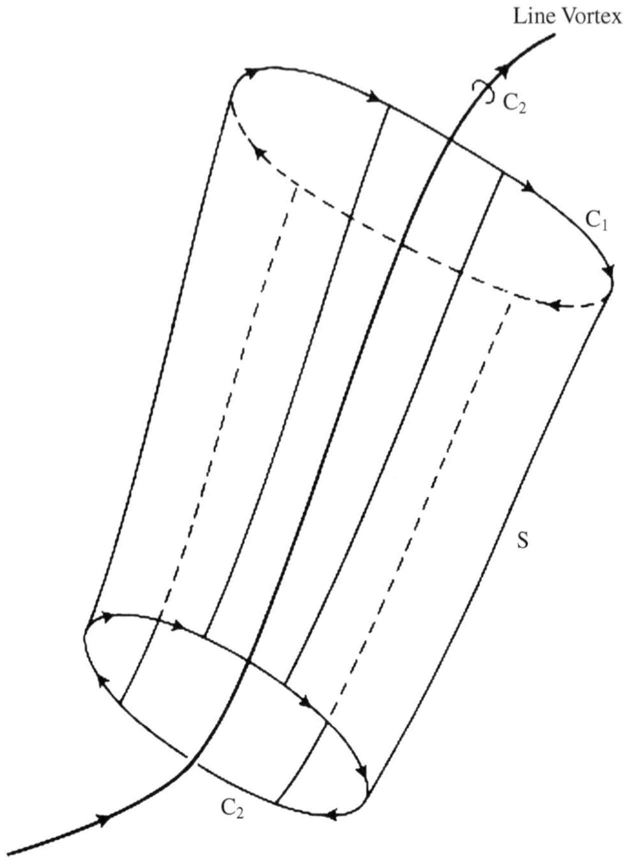

Figure 2.4: The circulations Γ_1 and Γ_2 around two different closed curves C_1 and C_2, that each embrace a line vortex exactly once in the same sense, are necessarily equal. This is proven by applying Stokes's theorem to a collar-shaped surface S whose boundary C consists of C_1 and C_2 taken in *opposite* sense. (From Lighthill [8], with permission from IMA.)

For the simplicity of the properties of line vortex and vortex sheet, it may, often, allow us to make calculations of flows in which all of the vorticity is concentrated in very thin regions. Such calculations use the simple equation $\nabla \times \mathbf{v} = 0$ for irrotational motion outside these regions.

2.14 Navier-Stokes equations of motion

In this section the equation of continuity and the momentum equations for viscous incompressible fluid are presented without detailed derivations. For such details, see the work of Lamb [6], Rosenhead [14], and Batchelor [2]. A more sophisticated derivation, using the full power of the tensor properties, is given by Jeffreys [5].

Using the advanced analysis of vector calculus in developing the orthogonal curvilinear coordinate systems, the equations of mass and momentum are derived in an elegant recent book by Rahman [12]. For advanced water wave problems, the reader is referred to Rahman [11].

It is assumed that throughout the motion of any element of fluid, its mass is conserved; hence, for incompressible flow, the volume of the fluid element must remain constant. This condition yields the equation of continuity. The momentum equations, which must be satisfied by the flow quantities at each point of the fluid, are deduced by applying Newton's second law of motion to the fluid which occupies an elementary volume.

2.14.1 Cartesian coordinates

In a Cartesian coordinate set of axes, the equation of continuity and the equations of motion can be written as

$$\frac{\partial u}{\partial x}+\frac{\partial v}{\partial y}+\frac{\partial w}{\partial z}=0$$

$$\begin{aligned}\frac{\partial u}{\partial t}+u\frac{\partial u}{\partial x}+v\frac{\partial u}{\partial y}+w\frac{\partial u}{\partial z}&=-\frac{1}{\rho}\frac{\partial p}{\partial x}+X+\nu\nabla^2 u\\ \frac{\partial v}{\partial t}+u\frac{\partial v}{\partial x}+v\frac{\partial v}{\partial y}+w\frac{\partial v}{\partial z}&=-\frac{1}{\rho}\frac{\partial p}{\partial y}+Y+\nu\nabla^2 v\\ \frac{\partial w}{\partial t}+u\frac{\partial w}{\partial x}+v\frac{\partial w}{\partial y}+w\frac{\partial w}{\partial z}&=-\frac{1}{\rho}\frac{\partial p}{\partial z}+Z+\nu\nabla^2 w,\end{aligned}\tag{2.102}$$

where ρ is the density, μ the dynamic viscosity and $\nu=\frac{\mu}{\rho}$ the kinematic viscosity of the fluid, and

$$\nabla^2=\frac{\partial^2}{\partial x^2}+\frac{\partial^2}{\partial y^2}+\frac{\partial^2}{\partial z^2}.\tag{2.103}$$

Here u, v, and w are the velocity components along the x, y, and z directions, and given by $\dot{x}=u$, $\dot{y}=v$, and $\dot{z}=w$.

2.14.2 Cylindrical polar coordinates

With the cylindrical polar coordinates (r,θ,z), where $x=r\ \cos\theta$, $y=r\ \sin\theta$, and $z=z$, the equation of continuity and the equations of motion can be written as

$$\frac{\partial v_r}{\partial r}+\frac{1}{r}v_r+\frac{1}{r}\frac{\partial v_\theta}{\partial \theta}+\frac{\partial v_z}{\partial z}=0$$

$$\begin{aligned}&\frac{\partial v_r}{\partial t}+v_r\frac{\partial v_r}{\partial r}+\frac{v_\theta}{r}\frac{\partial v_r}{\partial \theta}+v_z\frac{\partial v_r}{\partial z}-\frac{v_\theta^2}{r}\\ &\quad=-\frac{1}{\rho}\frac{\partial p}{\partial r}+\nu\left\{\nabla^2 v_r-\frac{v_r}{r^2}-\frac{2}{r^2}\frac{\partial v_\theta}{\partial \theta}\right\}+X_r\end{aligned}$$

$$\frac{\partial v_\theta}{\partial t} + v_r\frac{\partial v_\theta}{\partial r} + \frac{v_\theta}{r}\frac{\partial v_\theta}{\partial \theta} + v_z\frac{\partial v_\theta}{\partial z} + \frac{v_r v_\theta}{r} = -\frac{1}{\rho}\frac{\partial p}{r\partial \theta} + \nu\left\{\nabla^2 v_\theta + \frac{2}{r^2}\frac{\partial v_r}{\partial \theta} - \frac{v_\theta}{r^2}\right\} + X_\theta$$

$$\frac{\partial v_z}{\partial t} + v_r\frac{\partial v_z}{\partial r} + \frac{v_\theta}{r}\frac{\partial v_z}{\partial \theta} + v_z\frac{\partial v_z}{\partial z} = -\frac{1}{\rho}\frac{\partial p}{\partial z} + \nu\nabla^2 v_z + X_z \tag{2.104}$$

where

$$\nabla^2 = \frac{\partial^2}{\partial r^2} + \frac{1}{r}\frac{\partial}{\partial r} + \frac{1}{r^2}\frac{\partial^2}{\partial \theta^2} + \frac{\partial^2}{\partial z^2}. \tag{2.105}$$

The velocity components are given by $\dot{r} = v_r$, $\dot{\theta} = \frac{v_\theta}{r}$ and $\dot{z} = v_z$.

2.14.3 Spherical polar coordinates

With spherical polar coordinates (r, θ, ϕ) where $x = r\ \sin\theta\ \cos\phi$, $y = r\ \sin\theta\ \sin\phi$, and $z = r\ \cos\theta$, the equation of continuity and the momentum equations can be written as

$$\frac{\partial v_r}{\partial r} + \frac{2}{r}v_r + \frac{1}{r}\frac{\partial v_\theta}{\partial \theta} + \left(\frac{1}{r}\ \cot\theta\right)v_\theta + \frac{1}{r\ \sin\theta}\frac{\partial v_\phi}{\partial \phi} = 0$$

$$\frac{\partial v_r}{\partial t} + v_r\frac{\partial v_r}{\partial r} + \frac{v_\theta}{r}\frac{\partial v_r}{\partial \theta} + \frac{v_\phi}{r\ \sin\theta}\frac{\partial v_r}{\partial \phi} - \frac{v_\theta^2 + v_\phi^2}{r} = -\frac{1}{\rho}\frac{\partial p}{\partial r} + \nu\left\{\nabla^2 v_r - \frac{2v_r}{r^2} - \frac{2}{r^2}\frac{\partial v_\theta}{\partial \theta} - \frac{2v_\theta\cot\theta}{r^2} - \frac{2}{r^2\sin\theta}\frac{\partial v_\phi}{\partial \phi}\right\} + X_r$$

$$\frac{\partial v_\theta}{\partial t} + v_r\frac{\partial v_\theta}{\partial r} + \frac{v_\theta}{r}\frac{\partial v_\theta}{\partial \theta} + \frac{v_\phi}{r\ \sin\theta}\frac{\partial v_\theta}{\partial \phi} + \frac{v_r v_\theta}{r} - \frac{v_\phi^2\cot\theta}{r} = -\frac{1}{\rho}\frac{\partial p}{r\partial \theta} + \nu\left\{\nabla^2 v_\theta + \frac{2}{r^2}\frac{\partial v_r}{\partial \theta} - \frac{v_\theta}{r^2\sin^2\theta} - \frac{2\cos\theta}{r^2\sin^2\theta}\frac{\partial v_\phi}{\partial \phi}\right\} + X_\theta$$

$$\frac{\partial v_\phi}{\partial t} + v_r\frac{\partial v_\phi}{\partial r} + \frac{v_\theta}{r}\frac{\partial v_\phi}{\partial \theta} + \frac{v_\phi}{r\ \sin\theta}\frac{\partial v_\phi}{\partial \phi} + \frac{v_r v_\phi}{r} + \frac{v_\theta v_\phi\cot\theta}{r} = -\frac{1}{\rho}\frac{1}{r\ \sin\theta}\frac{\partial p}{\partial \phi} + \nu\left\{\nabla^2 v_\phi - \frac{v_\phi}{r^2\sin^2\theta} + \frac{2}{r^2\sin\theta}\frac{\partial v_r}{\partial \phi} + \frac{2\cos\theta}{r^2\sin^2\theta}\frac{\partial v_\theta}{\partial \phi}\right\} + X_\phi \tag{2.106}$$

where

$$\nabla^2 = \frac{1}{r^2}\frac{\partial}{\partial r}\left(r^2\frac{\partial}{\partial r}\right) + \frac{1}{r^2\sin\theta}\frac{\partial}{\partial \theta}\left(\sin\theta\frac{\partial}{\partial \theta}\right) + \frac{1}{r^2\sin^2\theta}\frac{\partial^2}{\partial \phi^2}. \tag{2.107}$$

The velocity components are given by $\dot{r} = v_r$, $\dot{\theta} = \frac{v_\theta}{r}$ and $\dot{\phi} = \frac{v_\phi}{r\sin\phi}$. Navier-Stokes equations reduce to Euler's equation of motion when $\nu = 0$, i.e. when the fluid is inviscid.

Example 2.1

If the position vector $\mathbf{r} = (x, y, z)$ has magnitude r, and $f(r)$ is any differentiable function, show that **grad** $f(r) = f'(r)r^{-1}\mathbf{r}$.

Proof

Using the gradient operator ∇ we have $\nabla f(r) = f'(r)\nabla r = f'(r)\frac{\mathbf{r}}{r} = f'(r)r^{-1}\mathbf{r}$. This is the required proof.

Example 2.2

If $\mathbf{v} = (2x^2y, xz^2 - y^3, xyz)$, calculate **div v**, **curl v** and **div curl v**.

Solution

Using the ∇ operator, we have

$$\begin{aligned}
\nabla \cdot \mathbf{v} &= \frac{\partial}{\partial x}(2x^2y) + \frac{\partial}{\partial y}(xz^2 - y^3) + \frac{\partial}{\partial z}(xyz) \\
&= 4xy - 3y^2 + xy = 5xy - 3y^2. \\
\mathbf{curl\ v} &= \mathbf{i}\left(\frac{\partial}{\partial y}(xyz) - \frac{\partial}{\partial z}(xz - y^3)\right) \\
&\quad + \mathbf{j}\left(\frac{\partial}{\partial z}(2x^2y) - \frac{\partial}{\partial x}(xyz)\right) \\
&\quad + \mathbf{k}\left(\frac{\partial}{\partial x}(xz - y^3) - \frac{\partial}{\partial y}(2x^2y)\right) \\
&= \mathbf{i}(xz - x) + \mathbf{j}(0 - yz) + \mathbf{k}(z - 2x^2). \\
\mathbf{div\ curl\ v} &= \frac{\partial}{\partial x}(xz - x) + \frac{\partial}{\partial y}(-yz) + \frac{\partial}{\partial z}(z - 2x^2) \\
&= z - 1 - z + 1 = 0.
\end{aligned}$$

This is true for any arbitrary vector field.

Example 2.3

Within a fluid, a solid sphere of radius a moves in a circle of radius c so that at time t the coordinates of its centre are $(0, c\cos\omega t, c\sin\omega t)$. After writing the equation of the sphere's surface in the form $F(x, y, z, t) = 0$, deduce that the fluid velocity adjacent to it satisfies the boundary condition

$$ux + v(y - c\ \cos\omega t) + w(z - c\ \sin\omega t) = c\,\omega(z\cos\omega t - y\sin\omega t).$$

Solution

If the solid boundary is at rest, the condition of zero normal flow through it can be written as

$$\mathbf{v} \cdot \mathbf{n} = 0.$$

But if $\mathbf{v}_S$ is the velocity of the solid, the relative velocity is $\mathbf{v} - \mathbf{v}_S$ and the boundary condition is therefore

$$(\mathbf{v} - \mathbf{v}_S) \cdot \mathbf{n} = 0.$$

A certain alternative form of the above equation, which is very often useful, involves an interesting application of the operator $\frac{D}{Dt} = \frac{\partial}{\partial t} + \mathbf{v} \cdot \nabla$. It can be applied if the geometrical equation of the solid surface in Cartesian coordinates x, y, z is known at each time t, in (say) the form

$$F(x, y, z, t) = 0.$$

We consider now the rate of change DF/Dt following a particle of fluid on the solid boundary. The condition $\mathbf{v} \cdot \mathbf{n} = 0$ means that the particles motion relative to that of the solid surface is purely tangential to the boundary (it has zero normal component). Thus, the particle moves along the solid surface and the value of $F(x, y, z, t)$ following the particle continues to be zero as specified by the equation $(\mathbf{v} - \mathbf{v}_S) \cdot \mathbf{n} = 0$. In short, there is zero rate of change of F for a particle at the surface:

$$0 = DF/Dt = \partial F/\partial t + \mathbf{v} \cdot \nabla F.$$

Although this condition looks very different to $\mathbf{v} \cdot \mathbf{n} = 0$, they are, in fact, mathematically equivalent. Thus we calculate the boundary condition using this condition as illustrated below.

In this problem,

$$F(x, y, z, t) = x^2 + (y - c\ \cos\omega t)^2 + (z - c\ \sin\omega t)^2 - a^2.$$

$$\mathbf{v} = \mathbf{i}u + \mathbf{j}v + \mathbf{k}w.$$

Hence to find the boundary condition we calculate:

$$\frac{\partial F}{\partial t} = 2c\omega(y \sin\omega t - z\cos\omega t)$$

$$\nabla\ F = \mathbf{i}(2x) + \mathbf{j}2(y - c\cos\omega t) + \mathbf{k}2(z - c\sin\omega t)$$

And, therefore, the boundary condition can be written as

$$ux + v(y - c\ \cos\omega t) + w(z - c\ \sin\omega t) = c\omega(z\cos\omega t - y\sin\omega t).$$

This is the required solution.

Example 2.4

Each particle of water in an open vessel is moving in a circular path about a vertical axis with a speed $q(r)$ depending only on the radius r of the path. Prove that the vorticity is a vertically directed vector with magnitude $q'(r) + r^{-1}q(r)$.

If the magnitude of the vorticity takes a constant value ω_0, for $r < a$ and zero for $r > a$, determine the value of $q(r)$ in both regions. Also, find the pressure from the Bernoulli's equation of motion taking gravity into account and deduce the shape of the free surface, showing it to be lower by a distance $(\omega_0^2 a^2/4g)$ at the centre of the vortex than it is in the undisturbed fluid.

Solution

The vorticty $\boldsymbol{\omega} = \nabla \times \mathbf{v}$ can be calculated using the circular cylindrical polar coordinate (r, θ, z) as follows:

$$\nabla \times \mathbf{v} = \frac{1}{r}\begin{vmatrix} \mathbf{r} & r\boldsymbol{\theta} & \mathbf{k} \\ \dfrac{\partial}{\partial r} & \dfrac{\partial}{\partial \theta} & \dfrac{\partial}{\partial z} \\ v_r & r v_\theta & v_z \end{vmatrix} \tag{2.108}$$

In this particular problem, $\mathbf{v} = (0, rq(r), 0)$, and so the vertical component of vorticity will survive and hence it is given by $\omega = \frac{1}{r}\frac{d}{dr}(rq(r))$.

We can write this differential equation in two regions. It is given that for the region $r < a$, $q(r)$ satisfies the differential equation

$$\frac{1}{r}\frac{d}{dr}(rq(r)) = \omega_0,$$

the solution of which is simply

$$q(r) = \frac{\omega_0 r}{2}.$$

Also for the region $r > a$, $q(r)$ satisfies the differential equation

$$\frac{1}{r}\frac{d}{dr}(rq(r)) = 0,$$

the solution of which is simply

$$q(r) = C/r,$$

where C is an arbitrary constant. This completes the solution of the first part of the problem.

For the second part of the problem, in the region $r < a$, we evaluate the pressure at any point (r, z) from Bernoulli's equation

$$p/\rho + gz + \frac{1}{2}q^2(r) = p_0/\rho.$$

Using the value of $q(r)$ in this equation we obtain the pressure at any point (r, z) as

$$p = -\rho g z - \frac{\rho}{2}\left(\frac{\omega_0^2 r^2}{4}\right) + p_0.$$

Now the pressure on the circumference of the circular path, i.e., at $r = a$ for any z is simply

$$p = -\rho g z - \frac{\rho}{2}\left(\frac{\omega_0^2 a^2}{4}\right) + p_0.$$

The pressure at $z = 0$ is given by $p = -\frac{\rho}{2}\left(\frac{\omega_0^2 a^2}{4}\right) + p_0$. But at the free surface $z = \eta$, $p = p_0$ and so the free surface elevation is given by

$$\eta = -\frac{1}{2}\left(\frac{\omega_0^2 a^2}{4g}\right).$$

This completes the solution of the problem.

For further information, the reader is referred to the work of Phillips, O.M. [10] and Reynolds, O. [13] as listed in the reference section.

2.15 Exercises

1. If $p = r = \sqrt{x^2 + y^2 + z^2}$ and $q = \frac{xy}{z^2}$, calculate ∇p and ∇q. Show that the scalar product $\nabla p \cdot \nabla q = 0$. Could this result be predicted in advance without separately calculating ∇p and ∇q?
2. Verify the Divergence Theorem by evaluating

$$\int\int\int_V \nabla \cdot \mathbf{v} dV$$

and

$$\int\int_S \mathbf{v} \cdot \mathbf{n} dS,$$

where V is the sphere $x^2 + y^2 + z^2 \leq a^2$ and

$$\mathbf{v} = (2xy^2 + 2xz^2, x^2 y, x^2 z).$$

[Hint: The spherical coordinates are: $x = r\sin\theta\cos\phi$, $y = r\sin\theta\sin\phi$, $z = r\cos\theta$. The elementary lengths are: dr, $rd\theta$, $r\sin\theta d\phi$. Therefore $\int\int\int_V \nabla \cdot \mathbf{v} dV = \frac{8}{5}\pi a^5 = \int\int_S \mathbf{v} \cdot \mathbf{n} dS$.]
3. In a certain fluid motion, the velocity field is $\mathbf{v} = (ax, by, cz)$, where a, b and c are constants. Show that the fluid density ρ takes the value $\rho = \rho_0 \exp[-(a + b + c)t]$ at time t, given that $\rho = \rho_0$ throughout the fluid at $t = 0$. [Hint: Consider the continuity equation $\frac{d\rho}{dt} + \rho\nabla \cdot \mathbf{v} = 0$.]

4. If $\mathbf{u}$ and $\mathbf{v}$ are any two vector fields, show that the divergence of the vector product takes the form

$$div(\mathbf{u} \times \mathbf{v}) = \mathbf{v} \cdot curl\mathbf{u} - \mathbf{u} \cdot curl\mathbf{v}.$$

By applying the Divergence Theorem to $\mathbf{u} \times \mathbf{v}$ where $\mathbf{v}$ is a constant vector, show that

$$\int\int\int_V curl\mathbf{u} dV = \int\int_S \mathbf{n} \times \mathbf{u} dS.$$

5. In a velocity field $(cy^2 + cz^2, 2cxy, -2cxz)$, where c is constant, show that the density of each fluid element is constant. Show that the vortexlines are all parallel straight lines, and that along any very thin vortextube the magnitude of the vorticity takes a constant value. At what points are the principal axes of rate of strain parallel to the coordinate axes?
6. Verify Stokes's theorem for a vector field with components

$$u = 0, \quad v = z \sin\theta, \quad w = r\cos\theta,$$

where the surface S is specified in cylindrical polar coordinates as

$$r = a, \quad 0 \leq \theta \leq \pi, \quad 0 \leq z \leq b.$$

7. For any fluid motion it is possible to consider that the particle of fluid whose position vector was $\mathbf{r}$ at time $t = 0$ and to ask where the particle so defined has moved to at a particular later time t. If its position vector takes the value $\mathbf{R}$ at time t then we regard $\mathbf{R}$ as a vector field in relation to its variation with the initial position $\mathbf{r}$ of the particle. With $\mathbf{R}$ so regarded as a vector field, and $\boldsymbol{\omega}$ taken as the vorticity of the particle at time $t = 0$, investigate the expression $\boldsymbol{\omega} \cdot \nabla\mathbf{R}$. Show how Helmholtz's theorem implies that this expression represents the particle's vorticity at the later time.
8. For the vector field $\mathbf{v}$ defined as

$$\mathbf{v} = (z - 2x/r, 2y - 3z - 2y/r, x - 3y - 2z/r),$$

where $r = \sqrt{x^2 + y^2 + z^2}$, prove that $\nabla \times \mathbf{v} = 0$ and find ϕ such that $\mathbf{v} = \nabla\phi$, in which ϕ is the velocity potential.
9. Prove that, for steady irrotational flow of a compressible fluid of nonuniform ρ, the velocity potential ϕ satisfies an equation of continuity

$$\nabla \cdot (\rho\nabla\phi) = 0.$$

Deduce that a steady irrotational flow of compressible fluid in a finite simply connected region with its boundaries at rest has zero kinetic energy, so that the fluid itself is at rest.
10. Calculate the distribution of temperature, which results from the diffusion of heat with the diffusivity $\alpha = \frac{k}{\rho c}$, after a cold body is immersed in hot liquid. A thin slice of the body, of unit area, at distances from the boundary between

z and $z + dz$, is gaining heat transferred across unit area at the rate $-k\frac{\partial T}{\partial z}$ and losing it at the rate equal to the same quantity with z replaced by $z + dz$, thus giving a net rate of gain $k(\frac{\partial^2 T}{\partial z^2})dz$. But the thin slice has mass ρdz so that the rate of increase of heat can be written $c(\rho dz)\frac{\partial T}{\partial t}$ where c is the specific heat. Deduce the differential equation for the temperature T and look for a solution in which T has the functional form

$$T = f(\eta) \text{ with } \eta = z(\alpha t)^{-\frac{1}{2}}.$$

Show that

$$f'(\eta) = C \exp\left(-\frac{1}{4}\eta^2\right),$$

where C is a constant, and deduce the solution which satisfies the boundary conditions

$$f = T_h \text{ at } \eta = 0 \text{ and } f \to T_c \text{ as } \eta \to \infty$$

in terms of the indefinite integral of the above exponential. [Solution is: $\frac{T_h - T}{T_h - T_c} = erf(\frac{\eta}{2}) = \frac{2}{\sqrt{\pi}} \int_0^{\eta/2} exp(-x^2)dx$.]

References

[1] Aris, R., *Vectors, Tensors and the Basic Equations of Fluid Dynamics*, Prentice-Hall: Englewood Cliffs, New Jersey, 1962.

[2] Batchelor, G.K., *An Introduction to Fluid Dynamics*, Cambridge University Press: Cambridge, 1967.

[3] Curle, N. and Davies, H., *Modern Fluid Dynamics*, Vol. 1, D. Van Nostrand: London, 1968.

[4] Eckart, C., The equations of motion of sea-water. (*The Sea*, ed. M.N. Hill, Vol. 1, pp. 31–42, New York: Interscience, 1962.

[5] Jeffreys, H., Tidal friction in shallow seas. *Philoso. Royal Soc., London*, **A211**, pp. 239–264, 1931.

[6] Lamb, H., *Hydrodynamics*, 6th edn, Dover: New York, 1945.

[7] Landau, L.D. and Lifshitz, E.M., *Fluid Mechanics*, Pergamon Press: London, 1959.

[8] Lighthill, J., *An Informal Introduction to Theoretical Fluid Mechanics*, Clarendon Press: Oxford, 1986.

[9] Milne-Thomson, L.M., *Theoretical Hydrodynamics*, 4th edn, The Macmillan Co.: New York, 1960.

[10] Phillips, O.M., *The Dynamics of the Upper Ocean*, Cambridge University Press: Cambridge, 1966.

[11] Rahman, M., *Water Waves: Relating Modern Theory to Advanced Engineering Applications*, Clarendon Press: Oxford, 1995.

[12] Rahman, M., *Advanced Vector Analysis for Scientists and Engineers*, WIT Press: Southampton, 2007.

[13] Reynolds, O., An experimental investigation of the circumstances which determine whether the motion of water shall be direct or sinuous and the law of resistance in parallel channels. *Phil. Trans.*, **A174**, pp. 935–982. Also, Scientific Papers, **2**, pp. 51–105, 1883.
[14] Rosenhead, L. (ed.), *Laminar Boundary Layers*, Oxford University Press: Oxford, 1963.
[15] Wylie, C.R. and Barrett, L.C., *Advanced Engineering Mathematics*, McGraw Hill: New York, 1982.

CHAPTER 3

Mechanics of viscous fluids

Albert Einstein

Albert Einstein (March 14, 1879, to April 18, 1955) was a German-born mathematician and theoretical physicist. He is best known for his theory of relativity and specially mass–energy equivalence, $E = mc^2$. Einstein received the 1921 Nobel Prize in Physics "for his services to Theoretical Physics, and especially for his discovery of the law of photoelectric effects". He was born in Ulm, Württemberg, Germany, on March 14, 1879. Einstein's many contributions to physics include his special theory of relativity, which reconciled mechanics with electromagnetism, and his general theory of relativity, which extended the principle of relativity to non-uniform motion, creating a new theory of gravitation. Works by Albert Einstein include more than fifty scientific papers and also non-scientific books. In 1999, Einstein was named Time Magazine's *"People of the Century", and a poll of prominent physicists named him the greatest physicist of all time. In popular culture the name "Einstein" has become synonymous with genius.*

3.1 Introduction

We have seen in the previous chapter that the motion of a fluid particle can be described by partial differential equations. In the study of fluid flow problems,

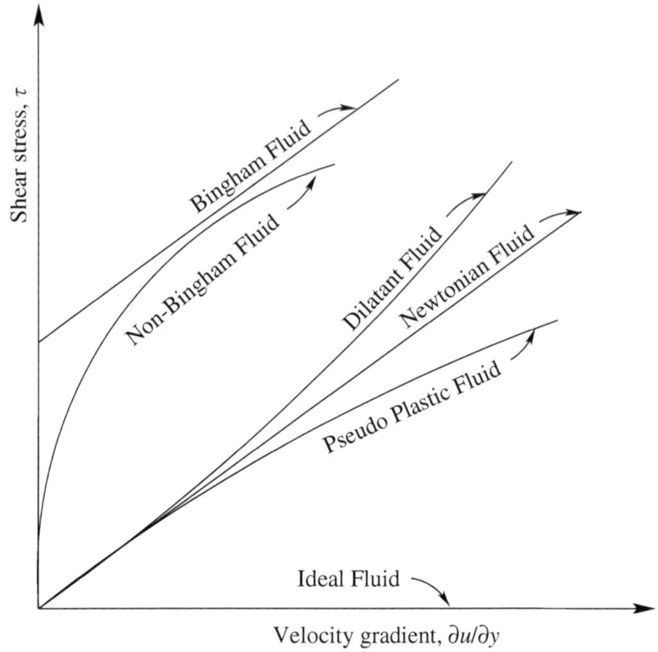

Figure 3.1: Rheological behaviour of various viscous fluids.

partial differential equations play a central role. We have seen, for example, how the velocity potential and the stream function, in fluid flow, are described by Laplace's equation. In this chapter we propose to deal with the mechanics of real fluids. Here again, the physics of the problems will be described by a set of partial differential equations. The solution techniques manifested in Rahman [9] will be employed here when they are warranted.

In this chapter, our study is directed to viscous incompressible fluids with the concept of Reynolds number. In chapter one, we have defined precisely the distinction between an ideal fluid and the viscous fluid. Ideal fluid theory assumes that if a surface is drawn in the fluid then the action exerted across the surface consists only of a normal pressure. In a real fluid, however, tangential stresses are possible, though in practice these are usually very small. The fluid property which causes these tangential stresses is known as viscosity.

For common fluids such as water and air, the shearing stress is proportional linearly to the variation of the velocity in the direction normal to the direction of the flow velocity. Such fluids are called Newtonian fluids. On the other hand, liquid which demonstrate a nonlinear behaviour or has a yield stress to deform is called a non-Newtonian fluid. These fluids are further classified as shown in Fig. 3.1 by the relationship between the shearing stress and the velocity gradient, i.e. a rheological diagram.

In many fluids, the coefficient of viscosity μ is very small, which is fundamentally the reason why viscous stresses are neglected in the ideal fluid-flow theory. In practice, the relative magnitude of viscous forces to pressure forces may be obtained as follows. The tangential stress produced by viscosity is directly proportional to the velocity gradient, i.e.,

$$\tau = \mu \frac{\partial u}{\partial y}.$$

The constant of proportionality, μ, depends entirely upon the physical properties of the fluid, and is called the coefficient of viscosity or dynamic viscosity, and τ is the stress or force per unit area in the x direction exerted by the faster moving fluid upon the slower moving fluid. If we consider a typical viscous stress of the above form and if U is at typical velocity and L a typical length in the flow, then the viscous forces are of order

$$\frac{\mu U}{L} \text{per unit area.}$$

In the same manner, a typical pressure force will be of the order

$$\rho U^2 \text{per unit area.}$$

The ratio of these two forces is accordingly

$$\frac{\text{typical pressure force}}{\text{typical viscous force}} = \frac{\rho U^2}{\mu U/L} = \frac{UL}{\nu}, \tag{3.1}$$

where $\nu = \frac{\mu}{\rho}$ is called the kinematic viscosity. The non-dimensional parameter

$$R = \frac{UL}{\nu}$$

is referred to as the Reynolds number. The viscosity μ and the kinematic viscosity ν are both functions of temperature and pressure. Typical values at atmospheric pressure and temperature of 20°C are

$$\mu = \left\{ \begin{array}{ll} 1.215 \times 10^{-5} \text{ lb/ft s} & \text{for air} \\ 0.675 \times 10^{-5} \text{ lb/ft s} & \text{for water} \end{array} \right\},$$

$$\nu = \left\{ \begin{array}{ll} 1.610 \times 10^{-4} \text{ ft}^2/\text{s} & \text{for air} \\ 1.084 \times 10^{-5} \text{ ft}^2/\text{s} & \text{for water} \end{array} \right\}.$$

In many practical illustrations it can be easily seen that for the medium of air and water the viscosity is so small that the Reynolds number becomes very large and in that situation, viscosity plays less important role in fluid flow problems. Nonetheless, viscosity can still play a very important role in many other flows. It is worth mentioning here that when the pressure gradients are either favourable or small, ideal fluid theory usually gives a reasonable picture of the behaviour of a real fluid. In the presence of large unfavourable pressure gradients, however, severe discrepancies can occur. With respect to large pressure gradients, we need a qualitative study of the flow near the boundary, and this region is usually referred to as boundary layer. We propose to investigate this important area of investigation in the next chapter.

3.2 Motion of a liquid in two-dimensions

Let us consider the problem of an infinitely long circular cylinder of radius a moving with velocity U perpendicular to its length in an infinite mass of liquid which is at rest at infinity. Let the origin be taken in the axis of the cylinder, and the axes of x, y in a plane perpendicular to its length. Further let the axis of x be in the direction of the velocity U. The motion, supposed originated from rest, will necessarily be irrotational, and the velocity potential ϕ will be single-valued. The fluid motion will not, of course, be steady but we can find the velocity potential at any given instant. As the motion is symmetrical we can write the solution of Laplace's equation $\nabla^2\phi = 0$ in cylindrical polar form as

$$\phi = \sum_{n=0}^{\infty} \left(A_n r^n + B_n r^{-n}\right) \cos n\theta \tag{3.2}$$

where the coefficient A_n and B_n must be found to satisfy the boundary conditions

$$\left(\frac{\partial \phi}{\partial r}\right)_{r=a} = U \cos\theta \tag{3.3}$$

the fluid velocity relative to the cylindrical surface $r = a$ is zero and

$$[\phi]_{r\to\infty} = 0 \tag{3.4}$$

the fluid being at rest at infinity. Condition (3.3) requires $n = 1$, and condition (3.4) requires $A_n = 0$. Therefore

$$\phi = \frac{B_1}{r} \cos\theta. \tag{3.5}$$

Using the surface boundary condition (3.3) we obtain the solution for ϕ as

$$\phi = -\frac{Ua^2}{r} \cos\theta \tag{3.6}$$

3.2.1 Pressure distribution

We know that the existence of the function ϕ such that

$$\nabla\phi = \mathbf{v}$$
$$\text{and} \quad \nabla^2\phi = 0,$$

satisfy the assumptions of irrotational motion and continuity. To complete the analysis of the problem, the solution of Euler's equation of motion is required. For the non-steady irrotational motion of incompressible inviscid fluid, we use Bernoulli's equation

$$\frac{p}{\rho} + \frac{1}{2}|\nabla\phi|^2 + \frac{\partial\phi}{\partial t} = f(t) \tag{3.7}$$

neglecting the external field. In the expression for the velocity potential of the fluid motion due to a moving cylinder,

$$\phi = -\frac{Ua^2}{r}\cos\theta,$$

both r and θ, being related to the centre of moving cylinder, are functions of t. Thus, in order to find the pressure on the cylinder it is required to find $\frac{\partial\phi}{\partial t}$. To do this it is convenient to write the expression for the velocity potential in the following form

$$\phi = -\frac{a^2}{r^2}(\mathbf{U}\cdot\mathbf{r}).$$

Thus differentiating the above equation with respect to time t partially, we obtain

$$\frac{\partial\phi}{\partial t} = -\frac{a^2}{r^2}\left(\frac{\partial\mathbf{U}}{\partial t}\cdot\mathbf{r}\right) - \frac{a^2}{r^2}\left(\mathbf{U}\cdot\frac{\partial\mathbf{r}}{\partial t}\right) + \frac{2a^2}{r^3}(\mathbf{U}\cdot\mathbf{r})\frac{\partial r}{\partial t} \tag{3.8}$$

If $\mathbf{R_0}$ is the position vector of the point P (fixed in space) and $\mathbf{R}$ is the position vector of the centre of the cylinder, both of which are referred to a fixed origin, then

$$\mathbf{r} = \mathbf{R_0} - \mathbf{R} \text{ and } r^2 = (\mathbf{r}\cdot\mathbf{r}),$$

so that

$$\frac{\partial\mathbf{r}}{\partial t} = -\frac{\partial\mathbf{R}}{\partial t} = -\mathbf{U}$$

$$\text{and } 2r\frac{\partial r}{\partial t} = 2\left(\mathbf{r}\cdot\frac{\partial\mathbf{r}}{\partial t}\right).$$

On substituting in (3.8) and expanding the scalar products we then obtain

$$\frac{\partial\phi}{\partial t} = -\frac{a^2}{r}\frac{dU}{dt}\cos\theta + \frac{a^2}{r^2}U^2 - 2\frac{a^2}{r^2}U^2\cos^2\theta \tag{3.9}$$

for an accelerating cylinder. The speed, $q = |\nabla\phi|$, at any point is given by

$$\begin{aligned} q^2 &= \left(\frac{\partial\phi}{\partial r}\right)^2 + \left(\frac{1}{r}\frac{\partial\phi}{\partial\theta}\right)^2 \\ &= \frac{a^4}{r^4}U^2 \end{aligned} \tag{3.10}$$

With the total pressure at infinity denoted by p_∞ the pressure at any point on the surface of the cylinder $r = a$ is then found, from the equation to be given by

$$\begin{aligned}\frac{p - p_\infty}{\rho} &= -\frac{\partial \phi}{\partial t} - \frac{1}{2}q^2 \\ &= a\frac{dU}{dt}\cos\theta - U^2 + 2U^2\cos^2\theta - \frac{1}{2}U^2 \\ &= a\frac{dU}{dt}\cos\theta - \frac{3}{2}U^2 + 2U^2\cos^2\theta \\ &= a\frac{dU}{dt}\cos\theta + \frac{1}{2}U^2(1 - 4\sin^2\theta). \end{aligned} \tag{3.11}$$

If U is uniformly constant and does not depend on t, then $\frac{dU}{dt} = 0$, and in that situation, we have

$$\frac{p - p_\infty}{\frac{1}{2}\rho U^2} = 1 - 4\sin^2\theta. \tag{3.12}$$

The above results are valid for the solid circular cylinder moving in the infinite liquid region when $r \geq a$. But for the hollow circular cylinder when the fluid motion is inside the cylinder $r \leq a$, then the velocity potential ϕ must satisfy the boundary condition $\frac{\partial \phi}{\partial r}|_{r=a} = U\cos\theta$ and ϕ must be finite at the centre of the cylinder $r = 0$. In this situation, then, the velocity potential must be

$$\phi = Ur\cos\theta \tag{3.13}$$

which is finite at the centre of the cylinder and at the same time it satisfies the body surface boundary condition. The speed of the fluid can be calculated as

$$\begin{aligned} q^2 &= \left(\frac{\partial \phi}{\partial r}\right)^2 + \left(\frac{1}{r}\frac{\partial \phi}{\partial \theta}\right)^2 \\ &= U^2\cos\theta + U^2\sin^2\theta \\ &= U^2 \end{aligned} \tag{3.14}$$

and hence $q = U$ same as the case when $r \geq a$. To determine the pressure inside and on the cylinder surface we use Bernoulli's equation. We know the velocity potential

$$\begin{aligned} \phi &= Ur\cos\theta \\ &= \mathbf{U}\cdot\mathbf{r} \end{aligned}$$

We calculate the $\frac{\partial \phi}{\partial t}$ as follows:

$$\begin{aligned} \frac{\partial \phi}{\partial t} &= \frac{d\mathbf{U}}{dt}\cdot\mathbf{r} + \mathbf{U}\cdot\frac{\partial \mathbf{r}}{\partial t} \\ &= \frac{dU}{dt}r\cos\theta - U^2 \end{aligned} \tag{3.15}$$

Therefore we have on the cylinder surface $r = a$

$$\frac{\partial \phi}{\partial t}|_{r=a} = a\frac{dU}{dt}\cos\theta - U^2. \tag{3.16}$$

With the total pressure at infinity denoted by p_∞ the pressure at any point on the surface of the cylinder $r = a$ is then found, from the equation to be given by

$$\begin{aligned}\frac{p - p_\infty}{\rho} &= -\frac{\partial \phi}{\partial t} - \frac{1}{2}q^2 \\ &= -a\frac{dU}{dt}\cos\theta + U^2 - \frac{1}{2}U^2 \\ &= -a\frac{dU}{dt}\cos\theta + \frac{1}{2}U^2.\end{aligned} \tag{3.17}$$

If U is constant, then $\frac{dU}{dt} = 0$, and so we have the dimensionless pressure as

$$\frac{p - p_\infty}{\frac{1}{2}\rho U^2} = 1.$$

The next important case may be to determine the velocity and pressure field when $0 < r < \infty$ the intermediate region of the fluid domain. This situation can be visualized by considering the fluid flow described by the complex potential defined by the uniform stream velocity U plus the singularity doublet. By the famous circle theorem we have the complex potential for a circular cylinder of radius a as follows:

$$W = U\left(z + \frac{a^2}{z}\right) \tag{3.18}$$

The velocity potential ϕ and the stream function ψ are, respectively, given by

$$\phi = U\left(r + \frac{a^2}{r}\right)\cos\theta$$

and

$$\psi = U\left(r - \frac{a^2}{r}\right)\sin\theta.$$

With these expressions we can determine the fluid velocity and pressure at any point in the fluid domain. With the complex potential $W = U(z + \frac{a^2}{z})$, the fluid velocity is obtained as

$$\begin{aligned} q &= \left|\frac{dW}{dz}\right| \\ &= U\left|\left(1-\frac{a^2}{z^2}\right)\right| \\ &= U\sqrt{\left(1-2\frac{a^2}{r^2}\cos 2\theta+\frac{a^4}{r^4}\right)} \\ &= U\sqrt{\left(1-\frac{a^2}{r^2}\right)^2+4\frac{a^2}{r^2}\sin^2\theta}. \end{aligned} \tag{3.19}$$

Also it is worth mentioning that $\frac{dW}{dz}=0$ predicts the stagnation points that are given by $z=\pm a$ where the fluid particle stops. But at $r=a$, i.e., on the surface of the cylinder (other than the two stagnation points)

$$q = 2U\sin\theta. \tag{3.20}$$

To determine the pressure on the circular cylinder with the given velocity potential $\phi = U(r+\frac{a^2}{r})\cos\theta = (1+\frac{a^2}{r^2})\mathbf{U}\cdot\mathbf{r}$, we need to find the values of $\frac{\partial\phi}{\partial t}$. So we proceed as before to obtain the expression of this term.

$$\begin{aligned} \frac{\partial\phi}{\partial t} &= \left(1+\frac{a^2}{r^2}\right)\frac{d\mathbf{U}}{dt}\cdot\mathbf{r}+\left(1+\frac{a^2}{r^2}\right)\mathbf{U}\cdot\frac{\partial\mathbf{r}}{\partial t}-\frac{2a^2}{r^3}\frac{\partial r}{\partial t}(\mathbf{U}\cdot\mathbf{r}) \\ &= \left(r+\frac{a^2}{r}\right)\frac{dU}{dt}\cos\theta-\left(1+\frac{a^2}{r^2}\right)U^2+\frac{2a^2}{r^2}U^2\cos^2\theta. \end{aligned}$$

Thus on the surface of the cylinder we have

$$\frac{\partial\phi}{\partial t}|_{r=a} = 2a\frac{dU}{dt}\cos\theta - 2U^2\sin^2\theta.$$

The pressure expression is then given by

$$\begin{aligned} \frac{p-p_\infty}{\rho} &= -\frac{\partial\phi}{\partial t}-\frac{1}{2}q^2 \\ &= -2a\frac{dU}{dt}\cos\theta. \end{aligned}$$

This result is good for the unsteady flow when $\frac{\partial\phi}{\partial t}\neq 0$. However, for steady flow, $\frac{\partial\phi}{\partial t}=0$, and so in this situation we have

$$\frac{p}{\rho}+\frac{1}{2}q^2 = C.$$

We know that at infinity, when $r \to \infty$, $q = U$ and $p = p_\infty$. Therefore $C = \frac{p_\infty}{\rho} + \frac{1}{2}U^2$. Thus

$$\frac{p - p_\infty}{\rho} = \frac{1}{2}U^2(1 - 4\sin^2\theta) \tag{3.21}$$

Which can be subsequently written in dimensionless form as

$$\frac{p - p_\infty}{\frac{1}{2}\rho U^2} = 1 - 4\sin^2\theta. \tag{3.22}$$

3.2.2 The drag force on the cylinder

The drag force on the cylinder may be obtained by integrating the resolved pressure force over the surface of the cylinder. Alternatively, this result may also be obtained by equating the rate of change of kinetic energy of the fluid to the work done by the fluid forces. As the motion is irrotational, we can write $\mathbf{v} = \nabla\phi$ and so $q^2 = |\nabla\phi|^2$. The kinetic energy T of the fluid of volume V bounded by the surface S is given by

$$T = \frac{1}{2}\int_V \rho q^2 dV \tag{3.23}$$

ρ being constant, as the fluid assumed to be incompressible. Equation (3.23) can be conveniently expressed using vector identity. We know that

$$\nabla \cdot \phi\nabla\phi = \nabla\phi \cdot \nabla\phi + \phi\nabla^2\phi.$$

As ϕ satisfies Laplace's equation $\nabla^2\phi = 0$, we have the simple expression for $q^2 = \nabla \cdot \phi\nabla\phi$. Then (3.23) can be written as

$$\begin{aligned} T &= \frac{\rho}{2}\int_V \nabla \cdot \phi\nabla\phi \\ &= \frac{\rho}{2}\int_s \phi\nabla\phi \cdot \mathbf{n}ds \\ &= \frac{\rho}{2}\int_s \phi\frac{\partial\phi}{\partial n}ds \end{aligned} \tag{3.24}$$

In obtaining this formula for T, we have used the divergence theorem of vector analysis. We know that the kinetic energy is given by $T = \frac{1}{2}\rho\int_s \phi\frac{\partial\phi}{\partial n}ds$, where $\partial\phi/\partial n$ is the specified velocity of the boundary along the outward normal from the fluid. This integration is to be performed along the circumference of the circle and hence the elementary length $ds = ad\theta$ and $\frac{\partial\phi}{\partial n} = -\frac{\partial\phi}{\partial r} = -U\frac{a^2}{r^2}\cos\theta$. Therefore $[\phi\frac{\partial\phi}{\partial n}]_{r=a} = (-Ua\cos\theta)(-U\cos\theta) = U^2 a\cos^2\theta$.

The kinetic energy of the fluid is

$$
\begin{aligned}
T_1 &= \frac{\rho}{2}\int_s \phi \frac{\partial \phi}{\partial n} ds \\
&= \frac{1}{2}\rho U^2 a^2 \int_0^{2\pi} \cos^2\theta d\theta \\
&= \frac{1}{2}[\pi \rho a^2]U^2 = \frac{M'}{2}U^2.
\end{aligned}
$$

where $M' = \rho\pi a^2$ is the mass of the liquid displaced by the cylinder. Usually $\frac{1}{2}M'$ is defined as the added mass of the cylinder due to the motion.

However, the kinetic energy produced by the moving cylinder is equal to $\frac{1}{2}MU^2$, where M is the mass of the cylinder. Therefore, the total kinetic energy is given by $T = \frac{1}{2}MU^2 + \frac{1}{2}M'U^2 = \frac{1}{2}(M + M')U^2$. The virtual mass of the cylinder is $M + M'$; thus the effect of the liquid is to increase the inertia of the sphere by half the mass of the liquid displaced.

If the total force (including the inertial field) component is X in the direction of U, then $\frac{dT}{dt} = \frac{d}{dt}[\frac{1}{2}MU^2 + \frac{1}{2}M'U^2]$ = rate at which work is being done = XU.

Therefore $M\frac{dU}{dt} + M'\frac{dU}{dt} = X$.

Writing this in the form

$$
M\frac{dU}{dt} = X - M'\frac{dU}{dt},
$$

we learn that the pressure of the fluid is equivalent to a force $-M'\frac{dU}{dt}$ per unit length in the direction of the motion. This vanishes when U is constant.

Note that the displaced liquid mass is equal to the mass of the cylinder itself.

3.3 Motion in an axially symmetric 3D-body

In two-dimensional motion, we have seen that the velocity potential ϕ, and the stream function ψ satisfy two-dimensional Laplace's equation. These two variables can be obtained as the real and imaginary parts of an analytic function, $W(z)$, which is known as the complex potential. Although no such development can be discussed in the case of general three-dimensional flow, a stream function may be defined for the case of axially symmetric flow, which represents the flux across any surface of revolution about the axis of symmetry. This stream function is attributed to Stokes [11]. Although the relationship between the stream function and the velocity potential does not satisfy the Cauchy-Riemann conditions, nevertheless the velocity potential ϕ satisfies the three-dimensional Laplace's equation. In spherical polar coordinates (r, θ, ω), Laplace's equation can be written as

$$
\frac{\partial}{\partial r}\left(r^2\frac{\partial \phi}{\partial r}\right) + \frac{1}{\sin\theta}\frac{\partial}{\partial \theta}\left(\sin\theta\frac{\partial \phi}{\partial \theta}\right) + \frac{1}{\sin^2\theta}\frac{\partial^2 \phi}{\partial \omega} = 0. \qquad (3.25)
$$

By using the method of separation of variables, the solution of ϕ can be written as

$$\phi = \sum_{n=0}^{\infty}\sum_{m=0}^{\infty}(A_n r^n + B_n r^{-n-1})P_n^m(\mu)\{C_m \cos m\omega + D_m \sin m\omega\} \tag{3.26}$$

where $P_n^m(\mu)$ is the solution of the ordinary differential equation known as the associated Legendre equation, namely

$$\frac{d}{d\mu}\left\{(1-\mu^2)\frac{dP}{d\mu}\right\} + \left\{n(n+1) - \frac{m^2}{1-\mu^2}\right\}P = 0 \tag{3.27}$$

where $\mu = \cos\theta$. For simplicity the superscript m and subscript n in the symbol P_n^m have been omitted.

For axi-symmetric flow about $\theta = 0$, the flow configuration in all axial planes, ω being constant, the velocity potential ϕ is independent of ω. This is equivalent to equating m to zero in (3.27), which then becomes Legendre's equation

$$\frac{d}{d\mu}\left\{(1-\mu^2)\frac{dP}{d\mu}\right\} + n(n+1)P = 0. \tag{3.28}$$

The solution of $P_n(\mu)$ can be obtained by means of Frobenius' method and exists in the following form:

$$P_n(\mu) = \sum_{r=0}^{p}\frac{(-1)^r(2n-2r)!}{2^n r!(n-r)!(n-2r)!}\mu^{n-2r} \tag{3.29}$$

where the integer p is $\frac{1}{2}n$ or $\frac{1}{2}(n-1)$, according as n is even or odd. Then the velocity potential ϕ in the axi-symmetric case is

$$\phi = \sum_{n=0}^{\infty}(A_n r^n + B_n r^{-n-1})P_n(cos\theta). \tag{3.30}$$

For instance, the motion induced by a sphere moving at the speed of $U(t)$ through an infinite fluid at rest, can be regarded as the axi-symmetric flow (see Fig. 3.1).

The velocity potential is given by (3.30). To determine the two constants, A_n and B_n, we need two boundary conditions, which are given by

$$\left(\frac{\partial\phi}{\partial r}\right)_{r=a} = U\cos\theta \tag{3.31}$$

and

$$(\phi)_{r\to\infty} = 0. \tag{3.32}$$

Using these two boundary conditions we see that (3.32) requires $A_n = 0$ and condition (3.31) yields

$$\left(\frac{\partial\phi}{\partial r}\right)_{r=a} = \sum_{n=0}^{\infty}[B_n(-n-1)r^{-n-2}P_n(\cos\theta)]_{r=a} = U\cos\theta \tag{3.33}$$

The first few Legendre polynomials ([2] and [14]) are

$$\begin{aligned}
P_0(\cos\theta) &= 1\\
P_1(\cos\theta) &= \cos\theta\\
P_2(\cos\theta) &= \frac{1}{2}(3\cos^2\theta - 1)\\
P_3(\cos\theta) &= \frac{1}{2}(5\cos^3\theta - 3\cos\theta)\\
P_4(\cos\theta) &= \frac{1}{8}(35\cos^4\theta - 30\cos^2\theta + 3).
\end{aligned}$$

3.3.1 Pressure distribution

In the following, we demonstrate to determine the pressure distribution on the surface of the sphere. Corresponding to n = 1, from (3.33) we have $-2B_1a^{-3} = U$, giving $B_1 = -\frac{1}{2}Ua^3$. Hence

$$\phi = -\frac{1}{2}Ua\left(\frac{a}{r}\right)^2\cos\theta. \tag{3.34}$$

The pressure distribution exerted on the surface of the sphere can be obtained from Bernoulli's equation

$$\frac{P}{\rho} + \frac{1}{2}(|\nabla\phi|^2) + \frac{\partial\phi}{\partial t} = C(t) \tag{3.35}$$

where $\phi = -\frac{1}{2}\frac{a^3}{r^2}U\cos\theta$ and

$$\frac{\partial\phi}{\partial t} = -\frac{1}{2}\frac{a^3}{r^2}\frac{dU}{dt}\cos\theta + \frac{a^3}{r^3}U\cos\theta\dot{r} + \frac{1}{2}\frac{a^3}{r^2}U\sin\theta\dot{\theta}. \tag{3.36}$$

If $\mathbf{R}_0$ is the position vector of the point P (fixed in space) and $\mathbf{R}$ is the position vector of the centre of the sphere, both of which are referred to a fixed origin (Fig. 3.1), then $\mathbf{r} = \mathbf{R}_0 - \mathbf{R}$. Thus $\dot{\mathbf{r}} = \dot{\mathbf{R}}_0 - \dot{\mathbf{R}} = -Ucos\theta$, $r\dot{\theta} = U\sin\theta$, and $\dot{\theta} = \frac{U\sin\theta}{r}$.

Substituting the above into the expression $\partial\phi/\partial t$, gives

$$\frac{\partial\phi}{\partial t} = -\frac{1}{2}\frac{a^3}{r^2}\cos\theta\frac{dU}{dt} + \frac{1}{2}\frac{a^3}{r^3}U^2 - \frac{3}{2}\frac{a^3}{r^3}U^2\cos^2\theta. \tag{3.37}$$

The speed at any point is given by $(|\nabla\phi|^2) = \left(\frac{\partial\phi}{\partial r}\right)^2 + \left(\frac{\partial\phi}{r\partial\theta}\right)^2 = \frac{U^2a^6}{r^6}\left(\cos^2\theta + \frac{1}{4}\sin^2\theta\right)$. From (3.37), the expression for the pressure is

$$\begin{aligned}
\frac{P}{\rho} = C(t) &+ \frac{1}{2}\frac{a^3}{r^2}\cos\theta\frac{dU}{dt} - \frac{1}{2}\frac{a^3}{r^3}U^2 + \frac{3}{2}\frac{a^3}{r^3}U^2\cos^2\theta\\
&- \frac{U^2a^6}{2r^6}\left(\cos^2\theta + \frac{1}{4}\sin^2\theta\right).
\end{aligned}$$

When $r \to \infty$, $p = p_\infty$ and hence $C(t) = \frac{P_\infty}{\rho}$. Thus the pressure force on the body of the sphere, r = a, is given by

$$\frac{P - P_\infty}{\rho} = \frac{1}{2}a\cos\theta\frac{dU}{dt} + \frac{1}{8}U^2(9\cos^2\theta - 5). \tag{3.38}$$

3.3.2 The drag force on the sphere

The drag force on the sphere can be obtained by integrating the resolved pressure force over the surface of the sphere. Alternatively, this result may be obtained by equating the rate of change of the kinetic energy of the fluid to the work done by the fluid forces. We know that the kinetic energy is given by $T = \frac{1}{2}\rho\int_s \phi\frac{\partial\phi}{\partial n}ds$, where $\partial\phi/\partial n$ is the specified velocity of the boundary along the outward normal from the fluid.

Referring to Fig. 3.2 we have $\delta s = (2\pi a\sin\theta)(a\delta\theta)$ and $\frac{\partial\phi}{\partial n} = -\frac{\partial\phi}{\partial r} = -\frac{Ua^3}{r^3}\cos\theta$. Therefore $[\phi\frac{\partial\phi}{\partial n}]_{r=a} = \frac{1}{2}aU^2\cos^2\theta$.

The kinetic energy of the fluid is

$$\begin{aligned}
T_1 &= \frac{\rho}{2}\int_s \phi\frac{\partial\phi}{\partial n}ds \\
&= \frac{\rho}{2}\int_0^\pi \left(\frac{1}{2}aU^2\cos^2\theta\right)(2\pi a^2\sin\theta d\theta) \\
&= \frac{\rho}{2}\pi a^3U^2\int_0^\pi \sin\theta\cos^2\theta d\theta \\
&= \frac{\rho}{2}\pi a^3U^2\frac{1}{3}(-\cos^3\theta)_0^\pi \\
&= \frac{\rho}{3}\pi a^3U^2 \\
&= \frac{1}{4}M'U^2,
\end{aligned}$$

where $M' = \frac{4}{3}\rho\pi a^3$ = mass of the liquid displaced by the sphere. Usually $\frac{1}{2}M'$ is defined as the added mass of the sphere due to the motion. However, the kinetic energy produced by the moving sphere is equal to $\frac{1}{2}MU^2$, where M is the mass of the sphere. Therefore, the total kinetic energy is given by $T = \frac{1}{2}MU^2 + \frac{1}{4}M'U^2 = \frac{1}{2}(M + \frac{1}{2}M')U^2$. The virtual mass of the sphere is $M + \frac{1}{2}M'$; thus the effect of the liquid is to increase the inertia of the sphere by half the mass of the liquid displaced. If the total force (including the inertial field) component is X in the direction of U, then $\frac{dT}{dt} = \frac{d}{dt}[\frac{1}{2}MU^2 + \frac{1}{4}M'U^2]$ = rate at which work is being done = XU. Therefore $M\frac{dU}{dt} + \frac{1}{2}M'\frac{dU}{dt} = X$.

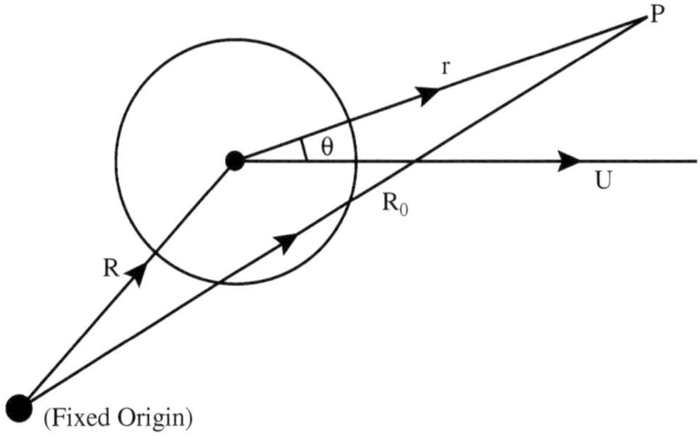

Figure 3.2: Moving sphere referred to fixed origin (from Rahman [9]).

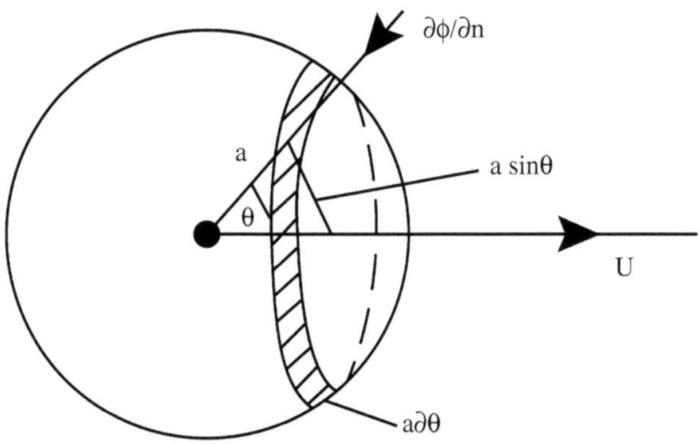

Figure 3.3: Drag force on a moving sphere (from Rahman [9]).

3.4 Distinction between ideal and real fluids

One of the simplest examples of the use of conformal mapping is that by which the two-dimensional inviscid incompressible flow past a circular cylinder (Fig. 3.3) is calculated. It is found that the flow pattern is perfectly symmetrical, as shown in the following figure, with stagnation points S at the front and rear of the cylinder. If the velocity and pressure far from the cylinder are U, p_0, respectively, then the velocity v and pressure p on the surface of the cylinder, at an angular distance θ radians from the forward stagnation point, are given by

$$q = 2U \sin\theta,$$

$$\frac{p - p_0}{\frac{1}{2}\rho U^2} = 1 - 4\sin^2\theta.$$

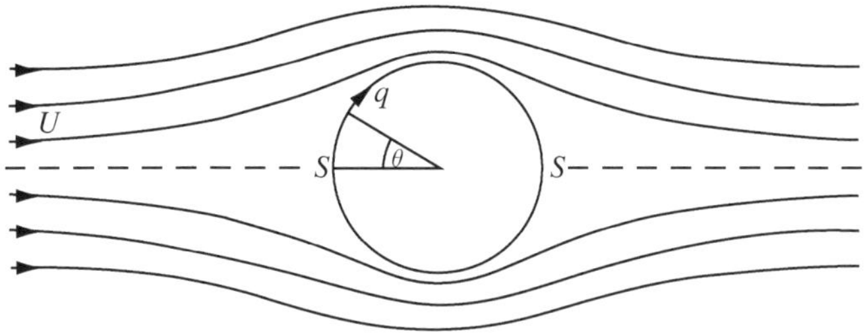

Figure 3.4: Inviscid flow past a circular cylinder (from Rahman [9]).

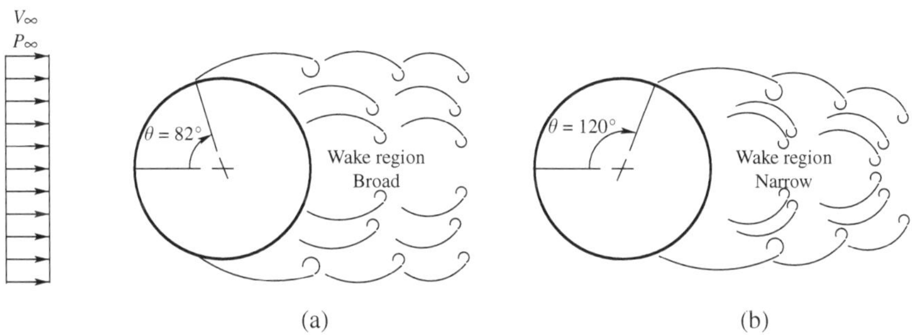

Figure 3.5: Real fluid flow around a circular cylinder: (a) Laminar flow; (b) Turbulent flow.

The velocity along the surface of the cylinder accordingly increases from zero at the forward stagnation point to a maximum of $2U$ at the position of maximum breadth of the cylinder, deceases symmetrically to zero at the rear stagnation point. The pressure, in an analogous way, has a maximum at the forward stagnation point, decreases to a minimum at $\theta = \pi/2$, and recovers fully to the same maximum at the rear stagnation point.

In Fig 3.5 real fluid flow around a circular cylinder is given for the case of laminar and turbulent flow. In Fig 3.4 a plot of this predicted values is shown, together with the pressure distributions measured in typical experimental flows at Reynolds number of about 2×10^5 and 7×10^5, when the flows are respectively subcritical and supercritical. It will be interesting to note that in each case the experimental pressure is in agreement with inviscid theory roughly the front quadrant of the cylinder, but that beyond this point the agreement becomes poorer, and that, in particular, the pressure rise at the rear of the cylinder is not obtained experimentally. A possible explanation of this observation is now given.

The approximation made in the ideal fluid theory may be described either physically or mathematically. Physically the assumptions, or simplification, is made that the action exerted across any surface in the fluid consists only of a normal pressure force acting whose magnitude is independent of the orientation of the surface. In a real fluid, tangential stresses are possible, though these are usually very

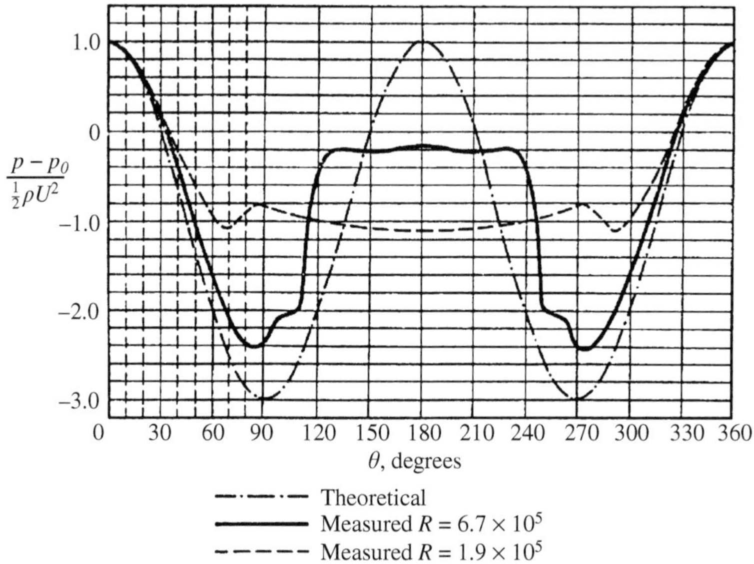

Figure 3.6: Inviscid theory and experiment for pressure on a circular cylinder (from [2]).

small. Mathematically, those terms in the full equations of motion which represent the tangential stresses are neglected. The neglected terms are those involving the higher derivatives of the velocity components, so that the order of the equations is reduced. As a result, fewer boundary conditions can be satisfied, and it is necessary to allow a velocity of slip at a solid boundary, whereas in a real fluid the boundary condition of zero velocity of slip is applied. More precisely, in the inviscid flow, a vortex sheet of appropriate strength is assumed to exist at the surface.

In view of the fact that ideal fluid theory allows a velocity of slip at a solid boundary, whereas in a real fluid no slip can exist, there must always be a region of slow- moving fluid close to the boundary, which is ignored in the ideal-fluid flow theory. This fluid will experience the pressure gradient along the surface, and will be affected considerably more than the faster-moving fluid further out. If the predicted pressure gradient is large enough and oppose the motion, then it is possible for the slow-moving fluid to be brought to rest, and even for a slow back- flow to be set up. The forward-moving fluid is then forced outwards to by-pass the back flow.

To sum up this qualitative discussion, we note that when the pressure gradients are either favourable or small, ideal- fluid theory usually gives a reasonable picture of the behaviour of a real fluid. In the presence of large unfavourable pressure gradient, however, severe discrepancies can occur, The question as to what constitutes a large pressure gradient in this context requires a qualitative investigation of the flow near to the wall. This region, usually referred to as the boundary layer, will be studied in the next chapter.

3.5 Drag forces in a real fluid

According to ideal-fluid theory, the over all force exerted by a solid body in a homogeneous fluid flow is exactly zero. On the other hand, it is observed experimentally that in general a body experiences a considerable drag, particularly if it is bluff. The reason why ideal-fluid theory predicts exactly zero drag may be illustrated by referring to the case of the circular cylinder, where the pressure distribution is symmetrical, excess pressure at the front of the body being exactly balanced by excess pressure at the rear. We may expect that if a body can be so designed that most of the predicted pressure recovery occurs, then the drag caused by the resultant pressure force will be close to the ideal-fluid value, that is close to zero. This is borne out by experiment, as we shall see shortly.

We know that the forces exerted on a body depends on the parameters of the flow. Let us suppose that flows of different fluids at different speeds past bodies of fixed shapes but variable scales. Then if L is the characteristic length of the body, and U is the typical fluid speed, the whole flow is determined by L, U, μ and ρ. It is worth noting that quantities as $\frac{p}{\rho U^2}$ are non-dimensional, that is they do not depend upon units used, although they may vary from position to position and from flow to flow. Thus $\frac{p}{\rho U^2}$ must depend on some non-dimensional combination of L, U, μ and ρ. Now the dimensions of these quantities are, respectively, L, LT^{-1}, $ML^{-1}T^{-1}$ and ML^{-3} where T is the characteristic time scale. Thus $L^\alpha \times U^\beta \times \mu^\gamma \times \rho^\delta$ has dimensions

$$(L^\alpha)\left(\frac{L}{T}\right)^\beta\left(\frac{M}{LT}\right)^\gamma\left(\frac{M}{L^3}\right)^\delta=\left(L^{\alpha+\beta-\gamma-3\delta}\right)\left(M^{\gamma+\delta}\right)\left(T^{-\beta-\gamma}\right).$$

This is non-dimensional if and only if

$$\begin{aligned}\alpha+\beta-\gamma-3\delta&=0\\ \gamma+\delta&=0\\ -\beta-\gamma&=0\end{aligned}$$

which implies that $\alpha=\beta=-\gamma=\delta$. We therefore conclude that the only non-dimensional combination of L, U, μ and ρ is $UL\mu^{-1}\rho=\frac{UL}{\nu}$ which may easily be recognized as the Reynolds number $R=\frac{UL}{\nu}$, and that the pressure p at the point is given by

$$p=\rho U^2F\left(\frac{x}{L},\frac{y}{L},\frac{z}{L},R\right).$$

Likewise the overall force exerted upon a body of surface area S in a given direction is of the form

$$\frac{F}{\frac{1}{2}\rho U^2S}=f(R).$$

Here the factor of one half in the denominator is a standard convention and S represents a typical area associated with body. The component of force in the direction of the undisturbed stream is called the drag force D, and a component right-angles to this is called the lift force L. The expressions

$$C_D = \frac{D}{\frac{1}{2}\rho U^2 S},$$

$$C_L = \frac{L}{\frac{1}{2}\rho U^2 S},$$

are called the drag coefficient and lift coefficient, respectively, and these are functions of R, which are in general determined by experiment.

3.6 Secondary flows

A further manifestation of real fluid effects may be seen in the phenomenon known as secondary flow. Consider, for example, the flow in a straight channel, representing an idealized river. Then according to ideal-fluid theory the flow across any section will be independent of position along the river; it will, in fact, be uniform. Suppose we now consider a channel having a slight bend, with straight section upstream and downstream of the bend. We may again assume that the flow is uniform in both the upstream and downstream sections, and that the fluid moving near to the outer wall of the bend will move faster than the fluid near to the inner wall. Accordingly, there must be a radial pressure gradient to balance the centrifugal force, the pressure being greatest at the outer wall and least at the inner wall. Consider, however, the fluid very close to the bed of the river. In a real fluid this will not be moving at the speed of the main body of fluid, but will be moving slowly, the fluid at the bed itself being at rest in order to satisfy the no-slip boundary condition. This fluid cannot withstand the radial pressure gradient imposed upon the fluid as a whole, and accordingly a secondary flow is set up, in which the fluid close to the bed of the river moves inwards. This motion clearly requires, by continuity, an upward motion close to the inner wall, an outward motion near the free surface and a downward motion close to the outer wall.

This secondary- flow phenomenon provides an explanation of the observation that a bend in a river loose materials are deposited near to the inside of the bend, whereas one might expect that they would be carried along in essentially straight paths and stick to the outer wall of the bend. The secondary flow ensures that material is removed from the outer side of the inner side of the bend, so that the bend becomes more and more pronounced.

3.7 Some exact solutions of Navier-Stokes equations

Navier-Stokes equations are inherently nonlinear because of the presence of convective term in the momentum equation which are nonlinear. The fundamental difficulty in solving the Navier-Stokes equation lies in the convective terms. There exist, of course, non-trivial flows in which the convective terms can be neglected, and these provide the simplest class of solutions of the equations of motion.

If we consider the components of velocity $v = 0$ and $w = 0$ except u, then immediately from the equation of continuity we infer that $\frac{\partial u}{\partial x} = 0$ which means u is independent of x. It follows that all the convective terms in the Navier-Stokes equations in Cartesian coordinates vanish. Neglecting the external forces X, Y and Z, the equations of motion can be written as

$$\begin{aligned}\frac{\partial u}{\partial t} &= -\frac{1}{\rho}\frac{\partial p}{\partial x} + \nu\left(\frac{\partial^2 u}{\partial y^2} + \frac{\partial^2 u}{\partial z^2}\right)\\ \frac{\partial p}{\partial y} &= 0\\ \frac{\partial p}{\partial z} &= 0.\end{aligned} \tag{3.39}$$

Since u is independent of x, we see that $\frac{\partial p}{\partial x}$ is a function of t alone. This pressure gradient can be prescribed by an arbitrary function of time t, then $u(y, z, t)$ is determined by solving the linear equation (3.39). This equation has a similarity with the two-dimensional heat conduction equation provided we treat the $(-\frac{1}{\rho}\frac{\partial p}{\partial x})$ as a uniform distribution of heat source. Thus the known solutions in the theory of heat conduction may be taken over differently and interpreted as the fluid flows.

It is clear that the flows to which this theory applies arc parallel to cylindrical structures whose generators are in the $x-$ direction. There are two main problems: (i) steady flows through the pipes of uniform cross-section with constant pressure gradient, and (ii) unsteady flows produced by the motion of a solid boundary in the $x-$ direction.

For steady state problem (i), $u(y, z)$ satisfies Poisson's equation

$$\frac{\partial^2 u}{\partial y^2} + \frac{\partial^2 u}{\partial z^2} = \frac{1}{\mu}\frac{dp}{dx} = -\frac{P}{\mu} \tag{3.40}$$

where $-\frac{dp}{dx} = P > 0$ and the boundary condition $u = 0$ at the wall of the pipe. It is clear that in all cases u may be expressed as

$$u = -\frac{1}{\mu}\frac{dp}{dx}f(y, z) = \frac{P}{\mu}f(y, z). \tag{3.41}$$

Thus (3.40) can be written in a simple form as

$$\frac{\partial^2 f}{\partial y^2} + \frac{\partial^2 f}{\partial z^2} = -1 \tag{3.42}$$

with the boundary condition $f = 0$ on the wall of the pipe, where $f(y, z)$ depends only in cross-sectional shape; similarly the mass flux takes the form $-M\frac{1}{\mu}\frac{dp}{dx} = M\frac{P}{\mu}$, where $M = \int\int_S \rho u(y, z)dydz$. The problem can be solved for several special shapes cross-section. We consider five different cases in the following for steady state flows.

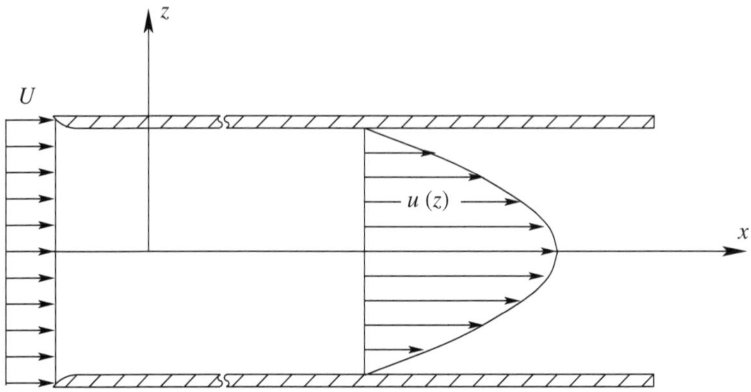

Figure 3.7: The fluid flow between two infinite parallel planes.

3.7.1 Steady flow between two-dimensional channel $-c \leq z \leq c$

We consider two infinite parallel planes to represent a two-dimensional channel, taken as $z = -c$, and $z = c$. Because the channel along the $y-$ direction is infinite, the differential equation (3.42) reduces to a simple form with the boundary conditions at $z = -c$ and at $z = c$.

$$\frac{d^2 f}{dz^2} = -1, \tag{3.43}$$

with boundary conditions $f(-c) = f(c) = 0$. The solution of (3.43) is simply $f(z) = -z^2/2 + Az + B$ where A and B are arbitrary constants. It can be easily seen that the two boundary conditions are satisfied if $A = 0$ and $B = c^2/2$. Hence the solution is $f(z) = \frac{1}{2}(c^2 - z^2)$. Therefore,

$$\begin{aligned} u(z) &= \frac{P}{\mu} f(z) \\ &= \frac{P}{2\mu}(c^2 - z^2) \end{aligned}$$

which is a parabolic profile as given in Fig. 3.7.

The mass–flux per unit width of the channel per unit time is obtained as

$$\begin{aligned} M &= \int_s \rho u(z) dz \\ &= \int_{-c}^{c} \rho \frac{P}{2\mu}(c^2 - z^2) dz \\ &= \frac{2\rho P}{3\mu} c^3. \end{aligned}$$

3.7.2 Steady flow through a circular section of radius c

Let us consider the steady flow through a circular cylinder of radius c as shown in Fig. 3.5. We transform the equation (3.42) into polar coordinates r, θ, such that $y = r\cos\theta$, and $z = r\sin\theta$, and note that the velocity u or f along the tube will be a function of r alone. Thus,

$$\nabla^2(f(r)) = \frac{1}{r}\frac{\partial}{\partial r}\left(r\frac{\partial f}{\partial r}\right) = -1,$$

which integrates to yield

$$r\frac{\partial f}{\partial r} = A - \frac{r^2}{2},$$

and hence

$$f(r) = A\ln r + B - \frac{r^2}{4}, \tag{3.44}$$

where A and B are arbitrary constants of integration. The constant A must be zero if the solution is to be physically acceptable along the axis $r = 0$, and B is then determined by condition that $f = 0$ when $r = c$,i.e., on the surface of the cylinder. Thus equation (3.44) becomes

$$f(r) = \frac{1}{4}(c^2 - r^2)$$

$$\text{and hence}\quad u(r) = \frac{P}{4\mu}(c^2 - r^2). \tag{3.45}$$

The velocity profile and the sectional description are given in Fig. 3.9.

From this velocity profile we may deduce the mass–flux per unit time passing any cross-section of the tube, namely

$$\begin{aligned} M &= \int\int_S (\rho u)dS \\ &= \int_{r=c}^{c}\int_{\theta=0}^{2\pi} \rho\frac{P}{4\mu}(c^2 - r^2)(rd\theta)dr \\ &= \frac{\pi\rho P}{2\mu}\int_0^c r(c^2 - r^2)dr \\ &= \frac{\pi\rho P c^4}{8\mu}. \end{aligned} \tag{3.46}$$

This result is known as Poiseuille's law. This law provides a basis of a method of measuring the viscosity of a fluid. Since the density of the fluid ρ and the

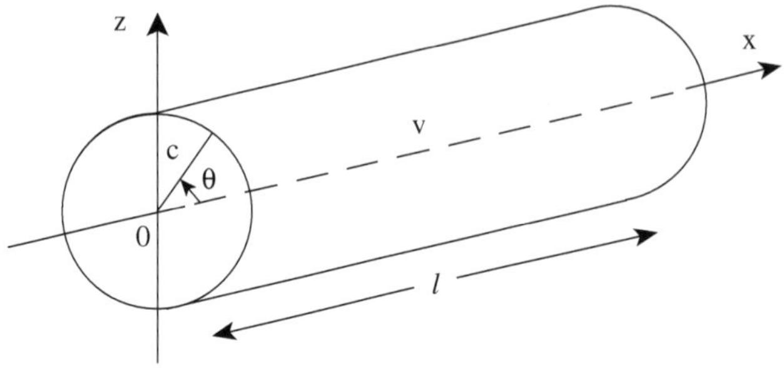

Figure 3.8: Steady flow through a circular cylinder.

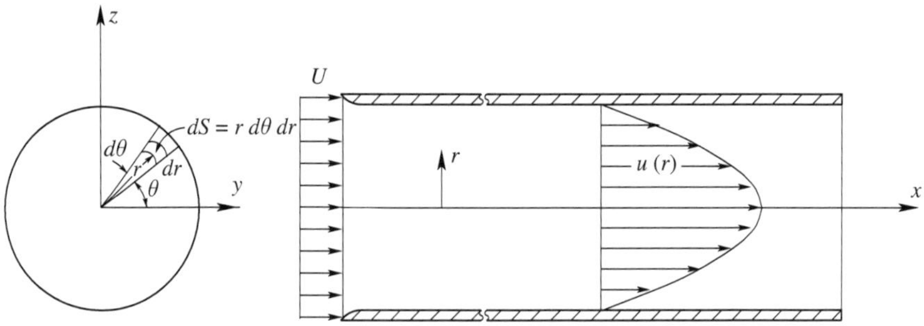

Figure 3.9: The fluid flow inside a circular cylinder.

radius of the tube c are known, we need only to measure the mass–flux and the pressure gradient to leave μ as the only unknown in equation (3.46), when μ can be calculated.

3.7.3 Steady flow through the annular region $b \leq r \leq c$

It is the extension of the previous problem, except that the boundary conditions are given by $(b) = u(c) = 0$, or $f(b) = f(c) = 0$. We have the general solution as given in (3.44), i.e.,

$$f(r) = A \ln r + B - \frac{r^2}{4}, \tag{3.47}$$

Using the boundary conditions we have the two algebraic equations

$$A \ln b + B = \frac{b^2}{4}$$

$$A \ln c + B = \frac{c^2}{4}$$

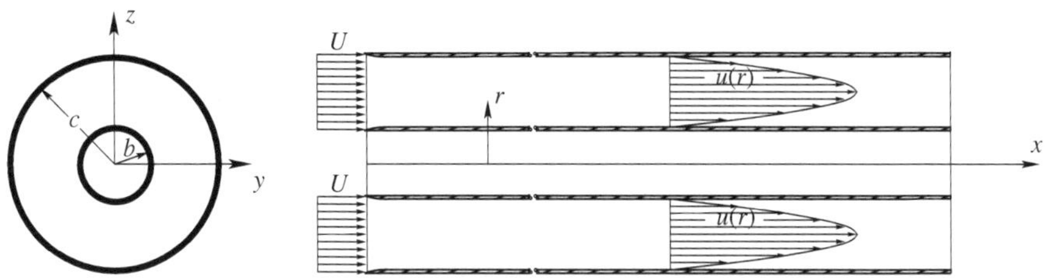

Figure 3.10: The fluid flow inside an annular region.

The solutions for A and B are given by

$$A = \frac{c^2 - b^2}{4\ln(c/b)}$$

$$B = \frac{b^2}{4} - \frac{(c^2 - b^2)\ln b}{4\ln(c/b)}$$

Using these two values in (3.47) yields the following after a little reduction

$$f(r) = \frac{1}{4}\left[b^2 - r^2 + \frac{c^2 - b^2}{\ln(c/b)}\ln(r/b)\right]$$

and subsequently, the expression for u is given by

$$u(r) = \frac{P}{4\mu}\left[b^2 - r^2 + \frac{c^2 - b^2}{\ln(c/b)}\ln(r/b)\right] \tag{3.48}$$

It can be easily verified that when $b \to 0$, (3.48) reduces to (3.45). The velocity profile is given in Fig. 3.10.

More generally, this result can be accomplished by using the following substitution. We write

$$\psi = u + \frac{P}{4\mu}(y^2 + z^2) \tag{3.49}$$

so that $\nabla^2\psi = \nabla^2 u + \frac{P}{\mu} = 0$. We note that ψ is a function of y and z only since u is, and the boundary condition is that $u = \psi - \frac{P}{4\mu}(y^2 + z^2) = 0$ on the boundary of the circular cylinder $y^2 + z^2 = c^2$, that is,

$$\psi = \frac{P}{4\mu}(y^2 + z^2) = \frac{Pc^2}{4\mu} \tag{3.50}$$

The solution of Laplace's equation $\nabla^2\psi = 0$ is simply $\psi = A\ln r + B$. But at $r = 0$ the solution must be finite. And hence $A = 0$. Therefore, using the boundary condition at $y^2 + z^2 = c^2$, yields $A = \frac{P}{4\mu}c^2$. Hence the solution for u is given by

$$\begin{aligned} u &= \psi - \frac{P}{4\mu}(y^2 + z^2) \\ &= \frac{P}{4\mu}\left\{c^2 - y^2 - z^2\right\} \\ &= \frac{P}{4\mu}(c^2 - r^2). \end{aligned}$$

3.7.4 Steady flow through an elliptic cylinder

Using the similar technique, we can determine the velocity profile u for an elliptic cylinder $\frac{y^2}{b^2} + \frac{z^2}{c^2} = 1$. We make the similar substitution as before

$$\psi = u + \frac{P}{2\mu}\frac{b^2c^2}{(b^2 + c^2)}\left(\frac{y^2}{b^2} + \frac{z^2}{c^2}\right)$$

such that

$$\nabla^2\psi = \nabla^2 u + \frac{P}{\mu} = 0.$$

The solution of Laplace's equation $\nabla^2\psi = 0$ is simply $\psi = B$. To determine B we use the boundary condition on the surface of the elliptical cylinder $y^2/b^2 + z^2/c^2 = 1$. This yields $B = \frac{P}{2\mu}\frac{b^2c^2}{(b^2+c^2)}$ because $u = 0$ on the surface. Hence the solution u can be written as

$$\begin{aligned} u &= \psi - \frac{P}{2\mu}\frac{b^2c^2}{(b^2 + c^2)}\left(\frac{y^2}{b^2} + \frac{z^2}{c^2}\right) \\ &= \frac{Pb^2c^2}{2\mu(b^2 + c^2)}(1 - y^2/b^2 - z^2/c^2) \end{aligned}$$

The profile of fluid velocity is depicted in Fig. 3.11.

The mass–flux across a section of the elliptic cylinder per unit time is

$$\begin{aligned} M &= \int\int_S (\rho u)dydz \\ &= (4)\frac{\rho Pb^2c^2}{2\mu(b^2 + c^2)}\int_{y=0}^{b}\int_{z=0}^{c\sqrt{1-y^2/b^2}}(1 - y^2/b^2 - z^2/c^2)dzdy \\ &= \frac{4c}{3}\frac{\rho Pb^2c^2}{\mu(b^2 + c^2)}\int_0^b (1 - y^2/b^2)^{3/2}dy \\ &= \frac{\pi\rho Pb^3c^3}{4\mu(b^2 + c^2)}. \end{aligned} \tag{3.51}$$

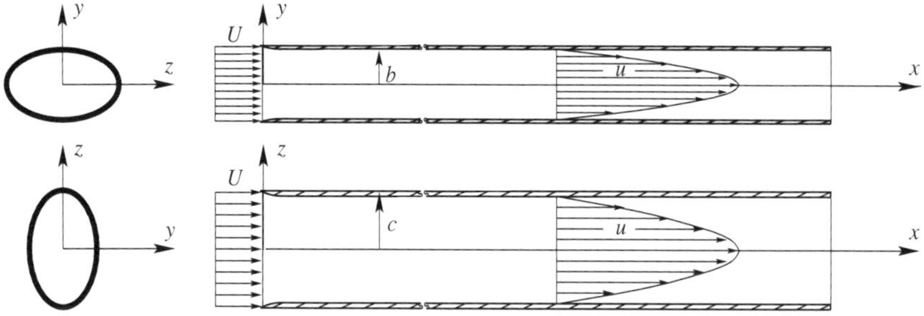

Figure 3.11: The fluid velocity profile through an elliptic cylinder.

For a circular cylinder of radius $\sqrt{bc}$, having the same cross-sectional area, the mass–flux is given by (3.46) as

$$M = \frac{\pi \rho P b^2 c^2}{8\mu},$$

and the ratio of these last two expressions yields the following interesting result that

$$\frac{M_c}{M_e} = \frac{b^2 + c^2}{2bc} = \frac{1}{2}\left(\frac{b}{c} + \frac{c}{b}\right),$$

where M_c = Mass–flux through circular cylinder, and M_e = Mass–flux through elliptic cylinder of same cross-sectional area.

3.7.5 Steady flow in a rectangular section

We consider a rectangular region $-b \le y \le b$ and $-c \le z \le c$. The flow is governed by Poisson's equation

$$\nabla^2 f = -1$$

with the zero boundary condition along the bounded region of the rectangular region. It can be easily verified that the particular solution of this problem is simply

$$f = \frac{b^2}{2} - \frac{y^2}{2},$$

and this solution satisfies the boundary conditions at $y = \pm b$ and at $z = \pm c$. Now to determine the general solution of the homogeneous equation, i.e., Laplace's equation

$$\nabla^2 f = 0,$$

we use the separation of variables method. The solution is as follows.

$$f(y, z) = (A \cos \lambda y + B \sin \lambda y)(C \cosh \lambda z + D \sinh \lambda z) \tag{3.52}$$

Using the boundary conditions at $y = \pm b$, it can be easily found that the eigenvalue

$$\lambda_n = \frac{(2n+1)\pi}{2b},$$

and hence the complete solution is given by

$$f(y,z) = \frac{b^2}{2} - \frac{y^2}{2} + \sum_{n=0}^{\infty}\left[A_n \cosh\frac{(2n+1)\pi z}{2b} + B_n \sinh\frac{(2n+1)\pi z}{2b}\right] \cos\frac{(2n+1)\pi y}{2b} \tag{3.53}$$

Using the boundary conditions at $z = \pm c$ we find that

$$A_n = -\frac{(-1)^n 6b^2}{\pi^3(2n+1)^3 \cosh\frac{(2n+1)\pi c}{2b}}$$

$$B_n = 0$$

Therefore, we have the solution for $f(y,z)$ as

$$f(y,z) = \frac{b^2}{2} - \frac{y^2}{2} - 2b^2\left(\frac{2}{\pi}\right)^3 \sum_{n=0}^{\infty} \frac{(-1)^n}{(2n+1)^3} \frac{\cosh(2n+1)\pi z/2b}{\cosh(2n+1)\pi c/2b} \cos(2n+1)\pi y/2b$$

and hence $u(y,z)$ is given by

$$u(y,z) = \frac{P}{\mu}\left[\frac{b^2}{2} - \frac{y^2}{2} - 2b^2\left(\frac{2}{\pi}\right)^3 \sum_{n=0}^{\infty} \frac{(-1)^n}{(2n+1)^3} \frac{\cosh(2n+1)\pi z/2b}{\cosh(2n+1)\pi c/2b} \cos(2n+1)\pi y/2b\right]$$

The mass–flux across the cross-section of the rectangular region per unit time is

$$\begin{aligned} M &= \int_{-c}^{c}\int_{-b}^{b} (\rho u)dydz \\ &= 4\frac{\rho P}{\mu}\int_0^c\int_0^b f(y,z)dydz \\ &= \frac{\rho P}{\mu}\left[\frac{4cb^3}{3} - 8b^4\left(\frac{2}{\pi}\right)^5 \sum_0^{\infty} \frac{1}{(2n+1)^5} \tanh(2n+1)\left(\frac{\pi c}{2b}\right)\right] \end{aligned} \tag{3.54}$$

3.7.6 Steady Couette flow between rotating cylinders

Let us consider now the viscous flow between two concentric, rotating circular cylinders of infinite length. Let r, θ, z be cylindrical polar coordinates and v_r, v_θ, v_z the corresponding components of velocity. Consider that the inner cylinder has angular velocity Ω_1 and radius r_1, and the outer cylinder has the angular velocity Ω_2 and radius r_2. Thus $r_2 > r_1$. We look for a solution in which

$$v_r = 0, \quad v_\theta = v_\theta(r), \quad v_z = 0, \quad p = p(r).$$

The equations of motion in cylindrical polar coordinates are given in the last section of Chapter 2 and for the special case under consideration, these equations become simply

$$\begin{aligned} -\frac{v_\theta^2}{r} &= -\frac{1}{\rho}\frac{\partial p}{\partial r} \\ 0 &= \nabla^2 v_\theta - \frac{v_\theta}{r^2} \end{aligned} \tag{3.55}$$

It is worth noting here that from the continuity equation we obtain $-\frac{1}{r}\frac{\partial v_\theta}{\partial \theta} = 0$, which implies that $v - \theta = v_\theta(r, z)$ but here we assume that v_θ is entirely a function of r alone. The second equation of (3.55) can be explicitly written as

$$\frac{d^2 v_\theta}{dr^2} + \frac{1}{r}\frac{dv_\theta}{dr} - \frac{v_\theta}{r^2} = 0.$$

This equation is of Euler-Cauchy type which can be transformed to a simple differential equation by the change of independent variable $r = e^\eta$ to the following form

$$\frac{d^2 v_\theta}{d\eta^2} - v_\theta = 0.$$

The solution of this equation is simply

$$\begin{aligned} v_\theta &= Ae^\eta + Be^{-\eta} \\ &= Ar + \frac{B}{r} \end{aligned} \tag{3.56}$$

The constants A and B are determined by boundary conditions that

$$v_\theta = r_1\Omega_1, \quad \text{when} \quad r = r_1,$$

and

$$v_\theta = r_2\Omega_2, \quad \text{when} \quad r = r_2.$$

The explicit form of these constants are

$$\begin{aligned} A &= \frac{r_1^2\Omega_1 - r_2^2\Omega_2}{r_1^2 - r_2^2} \\ B &= \frac{r_1^2 r_2^2(\Omega_2 - \Omega_1)}{r_1^2 - r_2^2} \end{aligned} \tag{3.57}$$

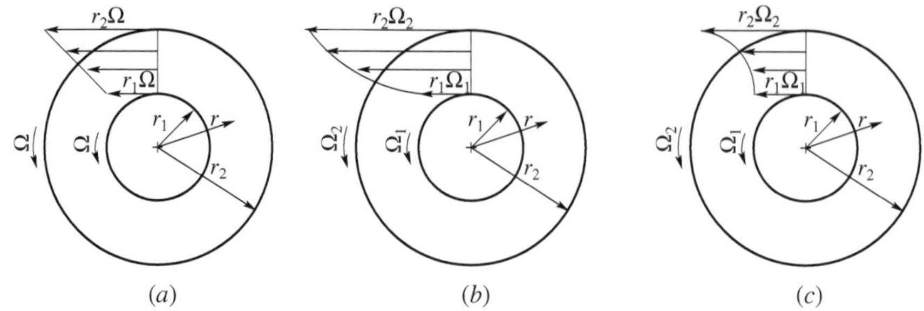

Figure 3.12: The fluid velocity profile between two rotating cylinder: (a) $\Omega_1 = \Omega_2 = \Omega$ (b) $\Omega_1 < \Omega_2$ (c) $\Omega_1 > \Omega_2$.

Thus the solution for v_θ is completely known with the known values of the constants A and B. The velocity profile is given for several cases in Fig. 3.12.

The pressure can be obtained from the first equation of (3.55), i.e.,

$$\frac{dp}{dr} = \rho \frac{v_\theta^2}{r},$$

which is given by after integration

$$p = \rho \left(A^2 \left(\frac{r^2}{2} \right) + 2AB(\ln r) - B^2 \left(\frac{1}{2r^2} \right) \right) + C.$$

The constant C depends on some characteristic pressure.

Remark

The relationship between stress and rate of strain in general orthogonal coordinates is derived in Love's *Mathematical Theory of Elasticity* [6]. For incompressible flow it is found that

$$p_{ij} = p\delta_{ij} - \mu \left\{ \frac{h_j}{h_i} \frac{\partial}{\partial x_i} \left(\frac{v_j}{h_j} \right) + \frac{h_i}{h_j} \frac{\partial}{\partial x_j} \left(\frac{v_i}{h_i} \right) + 2\delta_{ij} \sum_{j=1}^{3} \frac{v_j}{h_i h_i} \frac{\partial h_i}{\partial x_j} \right\} \tag{3.58}$$

where no summation is implied by a repeated suffix except where explicitly shown in the formula. For cylindrical polar coordinates, $h_1 = 1, h_2 = r$ and $h_3 = 1$. Using this information in the above equation (3.58) and taking into consideration that $v_r = 0$, $v_z = 0$, we can write the expression for stress-strain relationship as

$$p_{12} = -\mu r \frac{\partial}{\partial r} \left(\frac{1}{r} v_\theta \right).$$

Note that

$$\delta_{ij} = \begin{cases} 1 & i = j \\ 0 & i \neq j. \end{cases}$$

The force due to rotation of the cylinder exerted on the walls (usually known as *torque*) can be obtained from the following formula

$$\begin{aligned}
\tau &= \int\int_S p_{12} dS \\
&= \int_{r_1}^{r_2}\int_{\theta=0}^{2\pi} -\mu r\frac{\partial}{\partial r}\left(\frac{1}{r}v_\theta\right) r d\theta dr \\
&= 4\pi\mu B\ln\left(\frac{r_2}{r_1}\right)
\end{aligned}$$

If the inner cylinder is at rest, i.e., $\Omega_1 = 0$, then $B = \frac{r_1^2 r_2^2(\Omega_2)}{r_1^2 - r_2^2}$.

3.7.7 Steady flow between parallel planes

Let us consider two infinite parallel planes, taken as $y = 0$, and $y = h$, as in Fig. 3.7. Basically there are two problems to consider. Firstly, when both the planes are at rest, flow can be caused by an appropriate pressure gradient, as in the flow through a tube, this flow being called plane Poiseuille flow. Secondly, a flow without pressure gradient can be set up when one plane moves relative to the other, such a flow is called plane Couette flow, the cylindrical analogue having been previously considered. We shall not consider these cases separately.

With coordinates as in Fig 3.7, we consider that the plane $y = 0$ is at rest, and the plane $y = h$ moves with velocity U, the flow is caused by the pressure gradient $\frac{dp}{dx}$ parallel to the planes. In this situation, we assume that $u = u(y)$, $v = 0$, and $w = 0$. The continuity and momentum equations reduce to simple forms as

$$\begin{aligned}
\text{continuity equation:}\quad & \frac{\partial u}{\partial x} = 0 \\
\text{momentum equations:}\quad & 0 = -\frac{1}{\rho}\frac{dp}{dx} + \nu\frac{d^u}{dy^2} \\
& \frac{\partial p}{\partial y} = 0 \\
& \frac{\partial p}{\partial z} = 0.
\end{aligned}$$

The boundary value problem can be recast as follows:

$$\frac{d^2u}{dy^2} = \frac{1}{\mu}\left(\frac{dp}{dx}\right) \tag{3.59}$$

The boundary conditions are

$$\begin{aligned}
u &= 0 \quad \text{when} \quad y = 0 \\
u &= U \quad \text{when} \quad y = h.
\end{aligned} \tag{3.60}$$

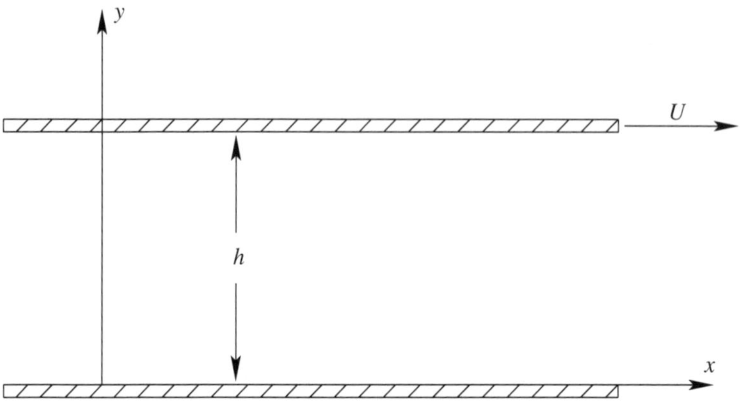

Figure 3.13: The coordinate for flow between two parallel plate.

It is worth noting that $\frac{1}{\mu}(\frac{dp}{dx})$ is a pure constant. The solution of this boundary value problem can be obtained at once and is given by

$$u = \frac{Uy}{h} + \frac{1}{\mu}\left(\frac{dp}{dx}\right) y(y-h). \tag{3.61}$$

The mass–flux per unit time and unit width z over planes perpendicular to the $x-$ axis is derived as

$$\begin{aligned} M &= \int_0^h (\rho u) dy \\ &= \rho \frac{U}{h} \int_0^h y dy + \rho \frac{1}{\mu}\left(\frac{dp}{dx}\right) \int_0^h y(y-h) dy \\ &= \frac{\rho}{2} U h - \frac{\rho}{12\mu}\left(\frac{dp}{dx}\right) h^3. \end{aligned} \tag{3.62}$$

The retarding force per unit area experienced by the moving plane is obtained as

$$\begin{aligned} \tau &= \mu \frac{du}{dy}|_{y=h} \\ &= \mu \frac{U}{h} + \frac{1}{2}\frac{dp}{dx} h \end{aligned} \tag{3.63}$$

3.8 Reynolds theory of lubrication

This generalized Couette flow is the basis for Reynolds theory of lubrication. To develop this theory, we consider a body, separated from a fixed plane by a thin layer of fluid, as shown in Fig 3.7. We assume that the region occupied by the fluid is approximately that between two parallel planes, and that the velocity profile at each

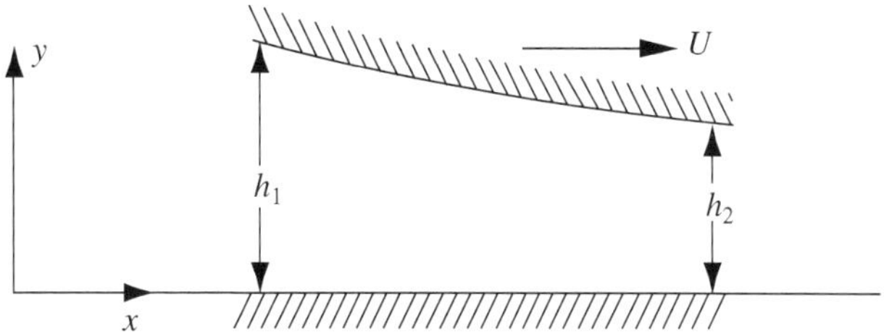

Figure 3.14: Reynolds lubrication theory (from [2]).

station is approximately of the form (3.61) with h and $\frac{dp}{dx}$ both depending slightly upon x. Thus

$$u = \frac{Uy}{h(x)} + \frac{1}{\mu}\left(\frac{dp}{dx}\right) y(y - h(x)). \tag{3.64}$$

The condition that the mass–flux in the x direction must be constant is then applied, and by (3.62) this gives

$$\frac{\rho}{2}Uh(x) - \frac{\rho}{12\mu}\left(\frac{dp}{dx}\right)h^3(x) = \text{constant} = \frac{\rho}{2}Uh_0, \quad \text{say}$$

where h_0 is an unknown constant to be determined later. The above equation may be written as

$$\frac{dp}{dx} = 6\mu U \frac{h - h_0}{h^3}. \tag{3.65}$$

Suppose we consider now the case when the upper body is almost a plane at an angle α to the lower plane. When α is very small and is positive then the stream is expanding that is $h_2 > h_1$ and when negative then the stream is contracting that is $h_2 < h_1$. We know that for very small α the slope $\frac{dh}{dx} = \tan\alpha \approx \alpha$ and (3.64) may be written as

$$\frac{dp}{dh} = \frac{dp}{dx}\frac{dx}{dh} = \frac{1}{\alpha}6\mu U \frac{h - h_0}{h^3}.$$

This equation integrates to yield

$$p = \frac{6\mu U}{\alpha}\left(-\frac{1}{h} + \frac{h_0}{2h^2}\right) + A \tag{3.66}$$

There are two unknown constants h_0 and A which are determined by the end conditions that $p = p_0$ when $h = h_1$ and also when $h = h_2$. Thus (3.66) must be expressed as

$$p - p_0 = \frac{6\mu U}{\alpha}\frac{(h - h_1)(h - h_2)}{(h_1 + h_2)h^2} \tag{3.67}$$

By usual procedure we can find the constants A and h_0 , and these constants are obtained as

$$A = p_0 + \frac{6\mu U}{\alpha(h_1 + h_2)}$$

$$h_0 = \frac{2h_1 h_2}{h_1 + h_2}$$

and the pressure equation can be obtained exactly as presented in (3.67).

In equation (3.67) we note that if $p - p_0$ is to be positive, yielding a thrust rather than a suction, then α must be negative, that is the stream must contract. Thus a necessary condition for lubrication is that the relative motion should tend to drag the fluid from the wider to the narrower part of the intervening space. It must be seen, by comparison of (3.66) and (3.67) that

$$h_0 = \frac{2h_1 h_2}{h_1 + h_2} \tag{3.68}$$

Having obtained the pressure distribution,we now calculate two very important factors namely the total thrust F, and the frictional resistance R experienced by the moving body. The normal stress exerted on the upper body is p_{22}, and by (3.58) this equals

$$p - 2\mu \frac{\partial v}{\partial y},$$

which is equal simply to p, since $v = 0$ in this flow. Thus the thrust (force) on the body, per unit width in the $z-$ direction, is

$$F = \int (p - p_0) dx$$

$$= \frac{1}{\alpha} \int_{h_1}^{h_2} (p - p_0) dh$$

and by (3.67) this gives

$$F = \frac{6\mu U}{\alpha^2 (h_1 + h_2)} \int_{h_1}^{h_2} \frac{(h - h_1)(h - h_2)}{h^2} dh$$

$$= \frac{6\mu U}{\alpha^2} \left\{ \frac{2(h_2 - h_1)}{h_1 + h_2} - \ln \left(\frac{h_2}{h_1} \right) \right\}$$

$$= \frac{6\mu U \ell^2}{(\lambda - 1)^2 h_2^2} \left\{ \ln \lambda - \frac{2(\lambda - 1)}{\lambda + 1} \right\} \tag{3.69}$$

where $\lambda = \frac{h_1}{h_2} > 1$, and ℓ is the length of the body. Here we have used the approximate value of $\alpha = \frac{dh}{dx} = \frac{h_1 - h_2}{\ell}$ such that $h(x) = h_1 + \alpha x$, and $h_2 = h_1 + \alpha \ell$.

In a similar manner the total frictional force R on the body can be obtained by integrating the expression given in (3.63) $\frac{\mu U}{h} + \frac{1}{2}\frac{dp}{dx}h$ with respect x

$$
\begin{aligned}
R &= \int \left\{ \frac{\mu U}{h} + \frac{1}{2}\frac{dp}{dx}h \right\} dx \\
&= \frac{1}{2}\int_{h_1}^{h_2} h\frac{dp}{dh}dh + \frac{\mu U}{\alpha}\int_{h_1}^{h_2}\frac{dh}{h} \\
&= \frac{3\mu U}{\alpha}\int_{h_1}^{h_2}\frac{h-h_0}{h^2}dh + \frac{\mu U}{\alpha}\int_{h_1}^{h_2}\frac{dh}{h}.
\end{aligned}
\tag{3.70}
$$

Integrating this equation and using the value of h_0 from (3.68) we obtain

$$
R = \frac{2\mu U \ell}{(\lambda - 1)h_2}\left\{ 2\ln\lambda - \frac{3(\lambda-1)}{\lambda+1} \right\}. \tag{3.71}
$$

For a maximum upward thrust we may deduce from (3.69) that $\lambda \approx 2, 2$,(this was found by Reynolds and confirmed by Rayleigh) so that the normal force (thrust) F and the frictional force(resistance) R on the body are given by

$$
F \approx 0.16\frac{\mu U \ell^2}{h_2^2}
$$

$$
\text{and} \quad \approx 0.75\frac{\mu U \ell}{h_2}.
$$

The ratio of resistance to thrust is simply

$$
\frac{R}{F} \approx 4.7\frac{h_2}{\ell}.
$$

Hence, by making h_2 small enough compared with ℓ, we can ensure a small frictional drag, i.e., good lubrication, and this is borne out experimentally.

Remark

The coordinate $\overline{x}$ of the centre of pressure is given by

$$
\begin{aligned}
F\overline{x} &= \int_0^{\ell} x(p-p_0)dx \\
&= \frac{1}{\alpha^2}\int_{h_1}^{h_2}(h-h_1)(p-p_0)dh \\
&= \frac{1}{\alpha^2}\left[\frac{(h-h_1)^2}{2}(p-p_0)\right]_{h_1}^{h_2} - \frac{1}{\alpha^2}\int_{h_1}^{h_2}\frac{(h-h_1)^2}{2}\left(\frac{dp}{dh}\right)dh
\end{aligned}
$$

$$= -\frac{1}{2\alpha^2}\int_{h_1}^{h_2}(h^2 - 2hh_1 + h_1^2)\left(\frac{dp}{dh}\right)dh$$

$$= \frac{1}{2\alpha^2}\int_{h_1}^{h_2}(2hh_1 - h_1^2)\left(\frac{dp}{dh}\right)dh - \frac{1}{2\alpha^2}\int_{h_1}^{h_2}h^2\left(\frac{dp}{dh}\right)dh$$

$$= \frac{\lambda\ell F}{\lambda - 1} - \frac{1}{2\alpha^2}\int_{h_1}^{h_2}h^2\left(\frac{dp}{dh}\right)dh$$

$$= \frac{\lambda\ell F}{\lambda 1} - \frac{3\mu U\ell^3}{(\lambda-1)^2h_2^2}\left(1 - \frac{2\lambda}{\lambda^2-1}\ln\lambda\right).$$

Hence the dimensionless coordinate of the centre of pressure is

$$\frac{\overline{x}}{\ell/2} = \frac{2\lambda}{\lambda^2-1} - \frac{\lambda^2 - 1 - 2\lambda\ln\lambda}{(\lambda^2-1)\ln\lambda - 2(\lambda-1)^2}.$$

When there is flow in the direction of y as well as x, we have

$$\int_0^h u dz = \frac{1}{2}hU - \frac{h^2}{12\mu}\frac{\partial p}{\partial x}$$

$$\int_0^h v dz = \frac{1}{2}hV - \frac{h^2}{12\mu}\frac{\partial p}{\partial y},$$

and the equation of continuity is

$$\frac{\partial}{\partial x}\int_0^h u dz + \frac{\partial}{\partial y}\int_0^h v dz = 0,$$

$$\text{or}\quad \frac{\partial}{\partial x}\left(h^2\frac{\partial p}{\partial x}\right) + \frac{\partial}{\partial y}\left(h^2\frac{\partial p}{\partial y}\right) = 6\mu\left\{\frac{\partial}{\partial x}(hU) + \frac{\partial}{\partial y}(hV)\right\}.$$

This can be applied to the case of rectangular block of finite dimensions sliding over a plane surface.

3.9 Steady flow due to a rotating circular disc

Let us consider that an infinite plane disc, rotating with angular velocity Ω in an otherwise unbounded fluid, at rest apart from the motion induced by the disc. We use cylindrical polar coordinates r, θ, z, with velocity components v_r, v_θ, v_z where $r = 0$ is the axis of rotation of the plane of the disc, $z = 0$ (Fig. 3.8). Then the equations of continuity and motion in cylindrical polar coordinates are as follows.

With cylindrical polar coordinates (r, θ, z) where $x = r\ \cos\theta$, $y = r\ \sin\theta$, and $z = z$, the equation of continuity and the equations of motion can be written as

$$\frac{\partial v_r}{\partial r} + \frac{1}{r}v_r + \frac{1}{r}\frac{\partial v_\theta}{\partial \theta} + \frac{\partial v_z}{\partial z} = 0 \qquad (3.72)$$

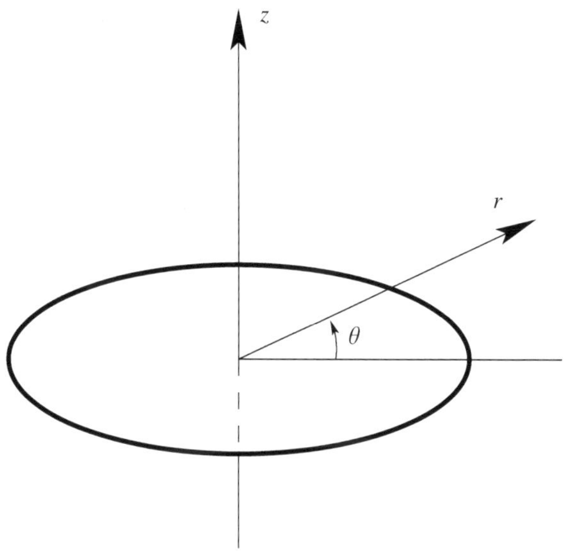

Figure 3.15: Coordinate system for rotating-disc flow.

$$\frac{\partial v_r}{\partial t} + v_r \frac{\partial v_r}{\partial r} + \frac{v_\theta}{r}\frac{\partial v_r}{\partial \theta} + v_z \frac{\partial v_r}{\partial z} - \frac{v_\theta^2}{r}$$
$$= -\frac{1}{\rho}\frac{\partial p}{\partial r} + \nu \left\{ \nabla^2 v_r - \frac{v_r}{r^2} - \frac{2}{r^2}\frac{\partial v_\theta}{\partial \theta} \right\} + X_r$$

$$\frac{\partial v_\theta}{\partial t} + v_r \frac{\partial v_\theta}{\partial r} + \frac{v_\theta}{r}\frac{\partial v_\theta}{\partial \theta} + v_z \frac{\partial v_\theta}{\partial z} + \frac{v_r v_\theta}{r}$$
$$= -\frac{1}{\rho}\frac{\partial p}{r\partial \theta} + \nu \left\{ \nabla^2 v_\theta + \frac{2}{r^2}\frac{\partial v_r}{\partial \theta} - \frac{v_\theta}{r^2} \right\} + X_\theta$$

$$\frac{\partial v_z}{\partial t} + v_r \frac{\partial v_z}{\partial r} + \frac{v_\theta}{r}\frac{\partial v_z}{\partial \theta} + v_z \frac{\partial v_z}{\partial z} = -\frac{1}{\rho}\frac{\partial p}{\partial z} + \nu \nabla^2 v_z + X_z \tag{3.73}$$

where

$$\nabla^2 = \frac{\partial^2}{\partial r^2} + \frac{1}{r}\frac{\partial}{\partial r} + \frac{1}{r^2}\frac{\partial^2}{\partial \theta^2} + \frac{\partial^2}{\partial z^2}. \tag{3.74}$$

The velocity components are given by $\dot{r} = v_r$, $\dot{\theta} = \frac{v_\theta}{r}$ and $\dot{z} = v_z$.

The boundary conditions for this problem are given by

$$v_r = 0, \quad v_\theta = r\Omega, \quad v_z = 0 \ \text{ when } \ z = 0,$$
$$v_r = v_\theta = 0 \ \text{ when } \ z \to \infty. \tag{3.75}$$

We look for a solution which is independent of θ and t. Then guided partly by the dimensional considerations and partly by the boundary conditions we consider the possibility that

$$v_\theta = \Omega r g(z),$$

and that v_r, v_z are each of the form

$$v_r = \Omega r f(z) \ \text{ or } \ (\nu\Omega)^{\frac{1}{2}} f(z),$$
$$v_z = \Omega r h(z) \ \text{ or } \ (\nu\Omega)^{\frac{1}{2}} h(z),$$

the argument being the Ωr and $(\nu\Omega)^{\frac{1}{2}}$ each have the dimension of velocity. By examining the equations of motion it is found that a solution of the type is possible only if we take

$$\begin{aligned} v_r &= \Omega r f(z), \\ v_\theta &= \Omega r g(z), \\ v_z &= \Omega r h(z), \\ \text{and } \ p &= \rho(\nu\Omega) p_1(z). \end{aligned} \tag{3.76}$$

Upon substituting these forms into (3.73) and (3.72) it is found, after some tedious algebraic reduction, that

$$\begin{aligned} 2f + \left(\frac{\nu}{\Omega}\right)^{\frac{1}{2}} h' &= 0, \\ f^2 - g^2 + \left(\frac{\nu}{\Omega}\right)^{\frac{1}{2}} f'h &= \frac{\nu}{\Omega} f'', \\ 2fg + \left(\frac{\nu}{\Omega}\right)^{\frac{1}{2}} g'h &= \frac{\nu}{\Omega} g'', \\ \text{and} \qquad hh' &= -p_1' + \left(\frac{\nu}{\Omega}\right)^{\frac{1}{2}} h'', \end{aligned} \tag{3.77}$$

where primes denote derivatives with respect to z. It is fairly straightforward to remove the constant coefficients involving $\frac{\nu}{\Omega}$ by writing

$$\eta = \left(\frac{\Omega}{\nu}\right)^{\frac{1}{2}} z,$$

such that

$$\begin{aligned} f(z) &= F(\eta), \\ g(z) &= G(\eta), \\ h(z) &= H(\eta), \\ p_1(z) &= P(\eta). \end{aligned}$$

Thus we have

$$\begin{aligned} v_r &= \Omega r F(\eta), \\ v_\theta &= \Omega r G(\eta), \\ v_z &= \Omega r H(\eta), \\ \text{and} \;\; p &= \rho(\nu\Omega) P(\eta). \end{aligned} \tag{3.78}$$

and equation (3.77) become

$$\begin{aligned} 2F + H' &= 0, \\ F^2 - G^2 + F'H &= F'', \\ 2FG + G'H &= G'', \\ \text{and} \qquad HH' &= -P' + H'', \end{aligned} \tag{3.79}$$

with boundary conditions

$$\begin{aligned} &F = 0, \quad G = 1, \quad H = 0 \;\; \text{when} \;\; \eta = 0, \\ &F \to 0, \quad G \to 0, \quad \text{when} \;\; \eta \to \infty. \end{aligned} \tag{3.80}$$

In effect the last of equations (3.79) may be regarded as an equation for P in terms of H.

A solution for small values of η

It is a set of nonlinear ordinary differential equations, and the solutions are obtained using the power series if η is small. Thus we write

$$\begin{aligned} F &= a_0 + a_1\eta + a_2\eta^2 + a_3\eta^3 + \cdots \\ G &= b_0 + b_1\eta + b_2\eta^2 + b_3\eta^3 + \cdots \\ H &= c_0 + c_1\eta + c_2\eta^2 + c_3\eta^3 + \cdots \end{aligned} \tag{3.81}$$

By virtue of the primary boundary conditions (3.80) at $\eta = 0$ we deduce that

$$a_0 = 0, \quad b_0 = 1, \quad c_0 = 0.$$

Further, by putting $\eta = 0$ in each of the equations (3.79) we deduce that

$$\begin{aligned} H'(0) &= 0 \\ F''(0) &= -1 \\ \text{and} \;\; G''(0) &= 0, \end{aligned}$$

so that

$$c_1 = 0, \quad a_2 = -\frac{1}{2}, \quad b_2 = 0.$$

By making use of equations (3.79) and the series expansion (3.81), we can obtain

$$a_3 = -\frac{1}{3}b_1, \quad b_3 = \frac{1}{3}a_1, \quad c_2 = -a_1, \quad c_3 = frac13.$$

By making use of the values of the coefficients a_n, b_n, c_n, so far obtained, we deduce that

$$\begin{aligned} F &= a_1\eta - \frac{1}{2}\eta^2 - \frac{1}{3}b_1\eta^3 = \cdots \\ G &= 1 + b_1\eta + \frac{1}{3}a_1\eta^3 \cdots \\ H &= -a_1\eta^2 + \frac{1}{3}\eta^3 \cdots \end{aligned} \tag{3.82}$$

where a_1 and b_1 alone are as yet unknown.

A solution for large values of η

We now look for a solution for large values of η. We note, first of all, that by virtue of the primary boundary conditions (3.80), the equations (3.79) may be approximated when η is large by

$$\begin{aligned} F'H_\infty &= F'' \\ \text{and} \quad G'H_\infty &= G'' \end{aligned}$$

Thus F' and G', be integrated, are both proportional to exp$(H_\infty\eta$, as H_∞ must be negative for consistency. If we write $H(\infty) = H_\infty = -c$, then for large η we have

$$\begin{aligned} F(\eta) &\propto e^{-c\eta} \\ G(\eta) &\propto e^{-c\eta} \\ \text{and} \quad H(\eta) &\to -c \end{aligned}$$

We may therefore reasonably look for a solution of the form

$$\begin{aligned} F &= A_1e^{-c\eta} + A_2e^{-2c\eta} + \cdots \\ G &= B_1e^{-c\eta} + B_2e^{-2c\eta} + \cdots \\ H &= -c + C_1e^{-c\eta} + C_2e^{-2c\eta} + \cdots \end{aligned}$$

By proceeding as before, the equations (3.79 may be shown to imply certain relationships between the unknown coefficients in these equations. In fact, it may be shown that

$$\begin{aligned} F &= A_1e^{-c\eta} - \frac{A_1^2 + B_1^2}{2c^2}e^{-2c\eta} + \cdots \\ G &= B_1e^{-c\eta} + \cdots \\ \text{and} \quad H &= -c\frac{2A_1}{c}e^{-c\eta} - \frac{A_1^2 + B_1^2}{2c^2}e^{-2c\eta} + \cdots \end{aligned} \tag{3.83}$$

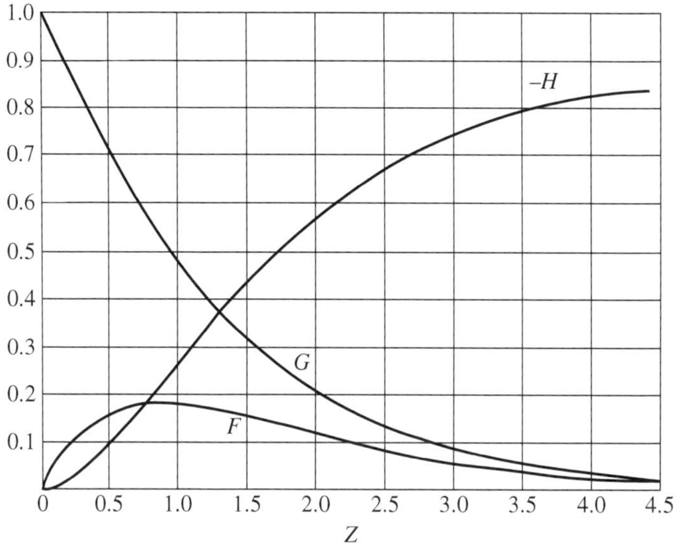

Figure 3.16: Functions $F(\eta)$, $G(\eta)$, $H(\eta)$ giving the velocity components in the flow produced by a rotating disc. Courtesy of Rosenhead [10] (from [2]).

where A_1, B_1 and c are to be determined.

We now choose A_1, B_1 c, a_1 and b_1, so that F, G, H, F' and G' are continuous where the expansion (3.81) and (3.83) are joined. By retaining a sufficient number of terms in each expansion any desired accuracy can be obtained, and the numerical results are

$$\begin{aligned} a_1 &= 0.510 \\ b_1 &= -0.616 \\ c &= 0.886 \\ A_1 &= 0.934 \\ B_1 &= 1.208. \end{aligned}$$

The functions F, G, H are depicted in Fig.3.9.

This figure shows that, for all practical purposes, F, G, and H have reached their limiting values when

$$\eta \approx 5\left(\frac{r}{\Omega}\right)^{\frac{1}{2}}.$$

It is worth noting therefore that the scale normal to the disc is proportional to $rR^{-\frac{1}{2}}$, where R, the Reynolds number, is equal to

$$R = \frac{\Omega r^2}{\nu} = \frac{(\Omega r)}{\nu} \tag{3.84}$$

For a finite disc, of radius a, we may calculate the retarding torque experienced by the disc, provided we ignore the effects of the edge, in the vicinity of which the pressure solution is not valid, since the boundary condition $v_\theta = \Omega r$ when $z = 0$ holds only on the surface of the disc, i.e., when $r \leq a$. The appropriate retarding shearing stress is p_{23}, and this is equal by (3.58)

$$\begin{aligned} p_{23} &= -\mu \left(\frac{\partial v_\theta}{\partial z} \right) \\ &= -\mu \Omega r \left(\frac{\Omega}{\nu} \right)^{\frac{1}{2}} G'(0). \end{aligned}$$

Thus the retarding torque on one side of the disc is

$$\begin{aligned} M &= \int_0^{2\pi} \int_0^a p_{23}(r d\theta) dr \\ &= 2\pi \int_0^a \left\{ -\mu \Omega r \left(\frac{\Omega}{\nu} \right)^{\frac{1}{2}} G'(0) \right\} r^2 dr \\ &= -\frac{1}{2}\pi G'(0)\mu \left(\frac{\Omega^3}{\nu} \right)^{\frac{1}{2}} a^4. \end{aligned} \tag{3.85}$$

This result may alternatively be expressed in terms of a non-dimensional "moment coefficient" by dividing by $\frac{1}{2}\rho a^2 \Omega^2 S$, where $S = \pi a^2$ is the surface area of the disc, and this yields

$$\begin{aligned} C_M &= -\left(\frac{\nu}{\Omega a^2} \right)^{\frac{1}{2}} G'(0) \\ &= 0.616 R_a^{-\frac{1}{2}} \end{aligned}$$

where $R_a = (\frac{(a\Omega)a}{\nu})$ is the Reynolds number based upon the radius a of the disc. The moment arising from the flow on two sides of the disc will, of course, be twice the value. That means $C_M = \frac{1.232}{R_a^{1/2}}$.

Remark

When the cross-section is not one of the special shapes that is for arbitrary cross-section for which an analytic solution can be found, the desired results may be obtained making certain measurements on soap films as liquids. For, if a soap film is stretched across a hole of a given shape and has a small excess pressure p on one side of it, then the displacement $X(y, z)$ satisfies

$$2T \left(\frac{\partial^2 X}{\partial y^2} + \frac{\partial^2 X}{\partial z^2} \right) + p = 0,$$

where T is the surface tension, together with the boundary condition $X = 0$ on the edge of the hole. Therefore,

$$X = \frac{p}{2T} f(y, z)$$

and is proportional to u. Thus the velocity distribution can be deduced from the measurement of the displacement X. The volume flux is proportional to $\int\int X dydz$ and this is found by measuring the total volume under the soap film. These measurements are much more easily made than direct measurements of the velocity in the fluid-flow problem or of the displacement in the torsion problem. The experimental technique is described by Taylor [12].

3.10 Some solutions of Navier-Stokes equations for unsteady flows

3.10.1 Flow due to motion of an infinite plate

In the previous section we have investigated many steady state fluid-flow problems with simple geometry. This section will be devoted to problems with unsteady flows. Perhaps the simplest case in this category is that of an infinite plate which, starting at $t = 0$, is moved in its own plane with constant velocity U_0 through fluid initially at rest. If the plate lies in the plane $z = 0$, the velocity $u(z, t)$ satisfies the first momentum equation

$$\frac{\partial u}{\partial t} = \nu \frac{\partial^2 u}{\partial z^2} \tag{3.86}$$

and the appropriate boundary and initial conditions are

$$\text{The boundary condition at } z = 0: \ u(0, t) = U_0 \ \ (t > 0) \tag{3.87}$$

$$\text{The boundary condition at } z \to \infty: \ u(\infty, t) = 0 \tag{3.88}$$

$$\text{The initial condition at } t = 0: \ u(z, 0) = 0 \tag{3.89}$$

It is worth mentioning here that the effect of the motion of the plate will diffuse outwards under the influence of viscosity. Since the plate is infinite we may reasonably assume that the extent of the diffusion will not depend on x, so that the solution will be of the form

$$\begin{aligned} u &= u(z, t) = f(\eta), \\ \text{where} \quad \eta &= \frac{z}{g(t)}, \\ v &= w = 0, \quad p = \text{constant} \end{aligned} \tag{3.90}$$

The equation (3.86) is exactly the same as the equation of heat conduction [1]. By analogy we may expect that the appropriate value of $g(t) = 2\sqrt{\nu t}$, and this can be shown in the following:

$$u_t = f'\eta_t = -f'\frac{z}{g^2}g' = -f'\eta\frac{g'}{g}$$

$$u_{zz} = \frac{f''}{g^2}$$

After substitution, and rearranging the terms we obtain

$$-\frac{f''}{\eta f'} = \frac{gg'}{\nu} = \text{constant.}$$

The form of $g(t)$ can be obtained as $g(t) = 2\sqrt{\nu t}$. Hence the solution for f is given by after integration

$$\begin{aligned} f(\eta) &= U_0\left\{1 - \frac{2}{\sqrt{\pi}}\int_0^{\eta} e^{-\xi^2}d\xi\right\} \\ &= U_0(1 - erf(\eta)) \\ &= U_0\ erfc(\eta) \end{aligned} \tag{3.91}$$

The procedure we have used here is known as the similarity technique. This problem can be very easily solved by using the Laplace transform method. We have after taking the Laplace transform of (3.86) and using the boundary and initial conditions

$$\begin{aligned} \frac{d^2}{dz^2}\mathcal{L}\{u\} - \frac{s}{\nu}\mathcal{L}\{u\} &= 0 \\ \text{at } z = 0,\quad \mathcal{L}\{u\} &= \frac{U_0}{s} \\ \text{at } z \to \infty\,,\quad \mathcal{L}\{u\} &\to 0. \end{aligned}$$

The solution of this set after taking the Laplace inverse yields

$$u(z,t) = U_0\ erfc\left(\frac{z}{2\sqrt{\nu t}}\right),$$

which is identical as before. Here $erf(\eta) = \frac{2}{\sqrt{\pi}}\int_0^{\eta} e^{-\xi^2}d\xi$, and $erfc(\eta) = \frac{2}{\sqrt{\pi}}\int_{\eta}^{\infty} e^{-\xi^2}d\xi$; and hence $erf(\eta) + erfc(\eta) = 1$.

3.10.2 Flow due to constant pressure gradient and motion of the plate

A second important solution for unsteady flow is that which applies when the flow is caused by constant pressure gradient in addition to motion of the infinite plate. In this situation, the differential equation is given by

$$\frac{\partial u}{\partial t} = -\frac{1}{\rho}\frac{\partial p}{\partial x} + \nu\frac{\partial^2 u}{\partial z^2} \tag{3.92}$$

and the appropriate boundary and initial conditions are

$$\text{The boundary condition at } z = 0: \; u(0, t) = U_0 \quad (t > 0)$$
$$\text{The boundary condition at } z \to \infty: \; |u(\infty, t)| = \text{finite}$$
$$\text{The initial condition at } t = 0: \; u(z, 0) = 0 \tag{3.93}$$

The solution can be effected by taking the Laplace transform of the differential equation and the boundary conditions

$$\frac{d^2}{dz^2}\mathcal{L}\{u\} - \frac{s}{\nu}\mathcal{L}\{u\} = -\frac{P}{\mu s} \tag{3.94}$$

$$z = 0, \quad \mathcal{L}\{u\} = \frac{U_0}{s}$$
$$z \to \infty, \quad \mathcal{L}\{u\} \to \text{finite}. \tag{3.95}$$

where $P = -\frac{\partial p}{\partial x}$.

The solution of (3.94) subject to the conditions (3.95) is given by

$$\mathcal{L}\{u\} = \frac{U_0}{s}e^{-\sqrt{s/\nu}z} - \frac{P}{\rho s^2}e^{-\sqrt{s/\nu}z} + \frac{P}{\rho s^2} \tag{3.96}$$

The inverse is obtained as

$$u(z, t) = U_0 \, erfc\left(\frac{z}{2\sqrt{\nu t}}\right) - \frac{P}{\rho}\left\{\left(t + \frac{z^2}{2\nu}\right) erfc\left(\frac{z}{2\sqrt{\nu t}}\right) - z\left(\frac{t}{\pi\nu}\right)^{\frac{1}{2}} e^{-z^2/4\nu t}\right\} + \frac{Pt}{\rho}.$$

It can be easily verified that this solution satisfies all the boundary and initial conditions. Note that $erfc(0) = 1$, and $erfc(\infty) = 0$. However, for large time, because of the presence of constant pressure gradient in the fluid flow, the solution will behave for large time t while z is fixed as

$$u \approx U_0 - \frac{P}{\rho}\left[\frac{z^2}{2\nu} - z\left(\frac{t}{\pi\nu}\right)^{\frac{1}{2}}\right].$$

In a similar manner, for large z, while t is kept fixed, the solution behaves like

$$u \approx \frac{Pt}{\rho}.$$

These observations are very interesting from the fluid-flow perspective.

3.10.3 Flow due to oscillation of the plate

A third important solution for unsteady flow is that which applies when the plate oscillates with a prescribed velocity

$$u(0, t) = U_0 e^{i\omega t}.$$

Suppose we consider the flow after a long time when a 'quasi-steady' oscillation has been set up. Thus, since (3.86) is linear, we may look for a solution satisfying the given boundary condition.

We try

$$u(z, t) = U_0 e^{i\omega t} f(z)$$

and substituting into the equation (3.86) yields, after simplification,

$$\begin{aligned} f'' &= \frac{i\omega}{\nu} f \\ &= \beta^2 f \end{aligned} \tag{3.97}$$

where $\beta = (\frac{\omega}{2\nu})^{\frac{1}{2}}(1+i) = k(1+i)$ say.

The solution of (3.97) is $\exp(\pm\beta z)$, and we choose the negative sign so that $f \to 0$ as $z \to \infty$. Thus we have

$$\begin{aligned} u &= U_0 \exp(i\omega t) \exp(-kz) \\ &= U_0 \exp(-kz) \exp(i(\omega t - kz)) \\ &= U_0 \exp(-kz) \{\cos(\omega t - kz) + i \sin(\omega t - kz)\} \end{aligned}$$

Therefore equating the real and imaginary parts of the above solution, we can write two solutions as

$$\begin{aligned} u_1 &= Re(u) = U_0 \exp(-kz) \cos(\omega t - kz) \\ u_2 &= Im(u) = U_0 \exp(-kz) \sin(\omega t - kz) \end{aligned}$$

Remark

For the motion of the infinite plate, the solution can be obtained for any prescribed variations of velocity of the plate with time. For the case of the plate oscillating periodically, that is, $u = U_0 \cos \omega t$ at $z = 0$, the solution satisfying the boundary condition is

$$u = U_0 e^{-kz} \cos(\omega t - kz), \quad k = \sqrt{\frac{\omega}{2\nu}}.$$

It represents waves spreading out from the plate with velocity

$$\frac{dz}{dt} = \left(\frac{\omega}{k}\right) = \sqrt{2\nu\omega},$$

and amplitude decaying exponentially with z. When ν is small the damping is heavy and the disturbance is then confined mainly to the thin boundary near the plate with thickness of order $\sqrt{\nu/\omega}$. The actual thickness is

$$k = \frac{2\pi}{\delta} = \sqrt{\frac{\omega}{2\nu}}.$$

And hence the thickness of the boundary layer is given by

$$\delta = 2\sqrt{2}\pi\sqrt{\nu/\omega}.$$

Therefore $\delta = O(\sqrt{\nu/\omega})$.

There are, of course, many other solutions of the heat conduction, which may be applied to fluid flows, but we do not attempt a complete survey in this book. We may consider some convective heat flow problems in the later sections. However, it should be pointed out that the solutions are not confined to the Cartesian form of equations. For example, in the cylindrical polar coordinates, we may consider, $v_r = v_z = 0$, and $v_\theta = v_\theta(r, t)$, $p =$ constant. Then v_θ satisfies the momentum equation

$$\frac{\partial v_\theta}{\partial t} = \nu\left\{\frac{\partial^2 v_\theta}{\partial r^2} + \frac{1}{r}\frac{\partial v_\theta}{\partial r} - \frac{v_\theta}{r^2}\right\}$$

and the vorticity

$$\omega_z = \frac{1}{r}\frac{\partial}{\partial r}(r v_\theta)$$

satisfying the diffusion equation

$$\frac{\partial \omega_z}{\partial t} = \nu\left\{\frac{\partial^2 \omega_z}{\partial r^2} + \frac{1}{r}\frac{\partial \omega_z}{\partial r}\right\}. \tag{3.98}$$

A well-known solution of this equation is

$$\omega_z = \frac{\Gamma}{4\pi\nu t}e^{-\frac{r^2}{4\nu t}},$$

in the application to fluid flow. It describes the dissolution of vortex filament, which is concentrated at the origin at $t = 0$, and Γ is the initial value of the circulation about the origin.

Another application of (3.98) is to the motion of fluid contained in or surrounding an infinite cylinder which starts to rotate. Equation (3.98) may be used to determine how the vorticity, which is initially concentrated at the surface of the cylinder, spreads out into the fluid. Outside the cylinder, $\omega_z \to 0$ as $t \to \infty$ and then $v_\theta \propto \frac{1}{r}$; inside the cylinder, ω_z tends to a constant value equal to twice the angular velocity of the cylinder, and the fluid rotates like a solid body.

Example 3.1

Find a solution of the vorticity equation

$$\frac{\partial \omega_z}{\partial t} = \nu \left\{ \frac{\partial^2 \omega_z}{\partial r^2} + \frac{1}{r} \frac{\partial \omega_z}{\partial r} \right\}.$$

in the form

$$\omega_z = f(t)g(\eta)$$

with

$$\eta = \frac{r}{2\sqrt{\nu t}}.$$

Verify that the velocity distribution by this approach is of the form

$$v_\theta = \frac{\Gamma}{2\pi r} \left\{ 1 - \exp\left(-\frac{r^2}{4\nu t} \right) \right\},$$

which represents a diffusing line vortex of strength Γ.

Solution

Given that $\omega = f(t)g(\eta)$ and hence we calculate

$$\begin{aligned} \frac{\partial \omega}{\partial t} &= f'(t)g(\eta) + f(t)g'(\eta)\eta_t \\ &= f'g - fg' \frac{r}{4\sqrt{\nu t^3}} \\ &= f'g - fg' \left(\frac{\eta}{2t} \right) \end{aligned}$$

Also we have

$$\begin{aligned} \frac{\partial \omega}{\partial r} &= f(t)g'(\eta)\eta_r = fg' \left(\frac{1}{2\sqrt{\nu t}} \right) = fg' \left(\frac{\eta}{r} \right) \\ \frac{\partial^2 \omega}{\partial r^2} &= fg'' \left(\frac{1}{4\nu t} \right) = fg'' \left(\frac{\eta^2}{r^2} \right) \end{aligned}$$

Thus substituting these expressions into the vortex equation yields

$$\begin{aligned} f'g - fg' \left(\frac{\eta}{2t} \right) &= \nu \left\{ fg'' \left(\frac{\eta^2}{r^2} \right) + fg' \left(\frac{\eta}{r^2} \right) \right\} \\ &= \nu f \left\{ g'' \left(\frac{\eta^2}{r^2} \right) + g' \left(\frac{\eta}{r^2} \right) \right\} \end{aligned}$$

Now after a little more reduction the variable can be separated as

$$(4t)\frac{f'}{f} = \frac{g''}{g} + \left(2\eta + \frac{1}{\eta} \right) \frac{g'}{g} = C(\text{a constant of separation}).$$

We have two ordinary differential equations to be solved.

$$f' = \left(\frac{C}{4t}\right) f$$

$$g'' + \left(2\eta + \frac{1}{\eta}\right) g' = Cg$$

The solution of the first equation can be written at once as

$$f(t) = Bt^{C/4};$$

and the solution of the second equation is assumed as an exponentially decaying solution as

$$g(\eta) = e^{-\eta^2}.$$

It is now easy to verify that these two solution will be valid provided $C = -4$, and hence the solution can be written as

$$w_z = \left(\frac{B}{t}\right) e^{-\eta^2} = \left(\frac{B}{t}\right) e^{-\frac{r^2}{4\nu t}}.$$

To determine the unknown constant B we evaluate the circulation around a huge circle using the vortex filament ω_z

$$\begin{aligned}
\Gamma &= \int_0^{2\pi} \int_0^{\infty} w_z(rd\theta)dr \\
&= \left(\frac{B}{t}\right)(2\pi) \int_0^{\infty} e^{-\frac{r^2}{4\nu t}} rdr \\
&= \left(\frac{B}{t}\right)(2\pi) \int_0^{\infty} \frac{1}{2} e^{-\frac{r^2}{4\nu t}} dr^2 \\
&= \left(\frac{B\pi}{t}\right)(4\nu t) \left[-e^{-\frac{r^2}{4\nu t}}\right]_0^{\infty} \\
&= B(4\pi\nu)
\end{aligned}$$

and hence $B = \frac{\Gamma}{4\nu\pi}$. Therefore

$$w_z = \left(\frac{B}{t}\right) e^{-\frac{r^2}{4\nu t}} = \frac{\Gamma}{4\pi\nu t} e^{-\frac{r^2}{4\nu t}}.$$

To determine the velocity component v_θ we use the relationship

$$w_z = \frac{1}{r}\frac{\partial}{\partial r}(r v_\theta).$$

Rearranging the terms, we have

$$\frac{\partial}{\partial r}(r v_\theta) = \frac{\Gamma}{4\pi\nu t} r e^{-\frac{r^2}{4\nu t}}.$$

After integration with respect to r from 0 to r yields

$$v_\theta = \frac{\Gamma}{2\pi r}\left\{1 - e^{-\frac{r^2}{4\nu t}}\right\}.$$

This is the required solution.

Remark

The circulation Γ of a velocity vector $\mathbf{v}$ around a closed circuit C, which surrounds a region S, is defined by

$$\begin{aligned}\Gamma &= \int_C \mathbf{v}\cdot d\mathbf{R}\\ &= \int\int_S \nabla\times\mathbf{v}\cdot\mathbf{n}dS \quad \text{(by Stokes's theorem)}\\ &= \int\int_S w_z\mathbf{k}\cdot\mathbf{k}dS \quad \text{(in this problem)}\\ &= \int_0^{2\pi}\int_0^{\infty} w_z(rd\theta)dr\end{aligned}$$

This vorticity problem can be successfully handled by using the Laplace transform method as described below. Taking the Laplace transform of the vorticity equation using the initial condition $w_z(r, 0) = 0$, we obtain an ordinary differential equation in terms of Laplace's parameter s and is given by

$$\frac{d^2}{dr^2}\mathcal{L}\{w_z\} + \frac{1}{r}\frac{d}{dr}\mathcal{L}\{w_z\} - \frac{s}{\nu}\mathcal{L}\{w_z\} = 0.$$

This equation, in essence, is a modified Bessel differential equation. The solution must be bounded in the interval $0 < r < \infty$. It has two solutions, namely, $I_0(\sqrt{s/\nu}r)$ and $K_0(\sqrt{s/\nu}r)$. We discard the $I_0(\sqrt{s/\nu}r)$ solution because it becomes infinite at ∞. So the acceptable solution is

$$\mathcal{L}\{w_z\} \sim K_0(\sqrt{s/\nu}r).$$

The inverse is given by

$$w_z = A\frac{1}{2t}e^{-r^2/4\nu t}$$

and so A must be equal to $\frac{\Gamma}{2\pi\nu}$ to comply with our previous result so that

$$w_z = \frac{\Gamma}{4\pi\nu t}e^{-r^2/4\nu t}.$$

3.11 Very slow motion

3.11.1 Stokes's flow using tensor calculus

We have already seen in the Navier-Stokes equations for in compressible flow,

$$\frac{\partial v_i}{\partial x_i} = 0,$$

$$\frac{\partial v_i}{\partial t} + v_j \frac{\partial v_i}{\partial x_j} = -\frac{1}{\rho}\frac{\partial p}{\partial x_i} + \nu \nabla^2 v_i,$$

the ratio of the inertia terms to the viscous terms is of the order $\frac{UL}{\nu}$, the Reynolds number. Thus, if the Reynolds number is very small, either because the velocity is very small, or because the scale of the flow is very small, or because the fluid is very viscous, then it should be possible to neglect the convective inertia terms which are usually nonlinear. When this is possible, the equations of motion take the following form

$$\frac{\partial v_i}{\partial x_i} = 0,$$

$$\frac{\partial v_i}{\partial t} = -\frac{1}{\rho}\frac{\partial p}{\partial x_i} + \nu \nabla^2 v_i. \tag{3.99}$$

These simplified equations are due to Stokes and, being linear, are easy to solve rather than the full Navier-Stokes equations. There are many physical situations for which the Reynolds number is vary small to allow us to apply the equations (3.99).

We now explore the method to determine the basic solutions of the equations (3.99) for the case of steady flow, for which they become

$$\frac{\partial v_i}{\partial x_i} = 0 \tag{3.100}$$

$$\text{and } \nabla^2 v_i = \frac{1}{\mu}\frac{\partial p}{\partial x_i}. \tag{3.101}$$

In much the same way as basic solutions can be found for the ideal inviscid fluid, for example, the source and the doublet, so we can look for a number of basic solutions of Stokes's equations. In fact it is fairly straightforward to obtain two families of solutions of these equations. By adding together suitable multiples of these solutions we can, in principle at any rate, obtain solutions with any required boundary conditions. For simple configurations, such as flow past a sphere, this procedure can be carried out in detail.

The first family of solutions may be obtained as follows. By differentiating equation (3.101) with respect to x_i , and using (3.100), we see

$$\frac{1}{\mu}\frac{\partial^2 p}{\partial x_i^2} = \nabla^2 \frac{\partial v_i}{\partial x_i} = 0,$$

that means

$$\nabla^2 p = 0.$$

Accordingly we may consider any function which satisfies Laplace's equation as the pressure p, and then obtain the velocity field by solution of (3.101). The obvious example is $p = \frac{1}{r}$, but this breaks down because the resulting velocities become infinite as $r \to \infty$, so the boundary conditions at infinity preclude this case. As a second example we take the three cases

$$p = \frac{2\mu x_j}{r^3} \quad j = 1, 2, 3, \tag{3.102}$$

for which the solution of (3.101) may be shown to be

$$v_i = \frac{1}{r}\delta_{ij} + \frac{x_i x_j}{r^3}, \quad i = 1, 2, 3.) \tag{3.103}$$

Addition to this family may be obtained by using any p whatsoever which satisfies Laplace's equation.

The second family of solutions to the basic equations is obtained by taking

$$v_i = \frac{\partial \phi}{\partial x_i}$$

where ϕ is any function which satisfies

$$\nabla \phi = 0.$$

All such solutions will automatically satisfy the continuity equation (3.100), since

$$\frac{\partial v_i}{\partial x_i} = \frac{\partial}{\partial x_i}\left(\frac{\partial \phi}{\partial x_i}\right) = \nabla^2 \phi = 0,$$

and will also satisfy (3.101) provided

$$\begin{aligned} \frac{1}{\mu}\frac{\partial p}{\partial x_i} &= \nabla^2\left(\frac{\partial \phi}{\partial x_i}\right) \\ &= \frac{\partial}{\partial x_i}(\nabla^2 \phi) \\ &= 0. \end{aligned}$$

In other words, the second family of solutions corresponds to uniform pressure.

As simple examples we consider

$$v_i = \frac{\partial}{\partial x_i}\left(\frac{1}{r}\right) = -\frac{x_i}{r^3}, \tag{3.104}$$

and the three solutions ($j = 1, 2, 3$)

$$\begin{aligned} v_i &= \frac{\partial}{\partial x_i}\left(\frac{x_i}{r^3}\right) \\ &= \frac{1}{r^3}\delta_{ij} - \frac{3x_i x_j}{r^5}. \end{aligned} \tag{3.105}$$

By adding together suitable multiples of the solutions (3.102) to (3.105) and of such further basic solutions as may be required, we can in principle obtain the solution of any problem. As an example we consider the case of flow past a sphere, for which the above solutions are adequate.

By adding arbitrary multiples of the above solutions to a uniform velocity field $(U, 0, 0)$ we can show that the Stokes solution for flow past a sphere, $r = a$, with uniform flow $(U, 0, 0)$ at infinity and satisfying the boundary conditions $v_1 = v_2 = v_3 = 0$ on the sphere, is

$$\begin{aligned} v_1 &= U\left\{1 - \frac{3a}{4r} - \frac{a^3}{4r^3} + \frac{3}{4}\frac{ax_1^2}{r^3}\left(\frac{a^2}{r^2} - 1\right)\right\}, \\ v_2 &= \frac{3}{4}U\frac{ax_1x_2}{r^3}\left(\frac{a^2}{r^2} - 1\right), \\ v_3 &= \frac{3}{4}U\frac{ax_1x_3}{r^3}\left(\frac{a^2}{r^2} - 1\right), \\ p &= -\frac{3\mu U a x_1}{2r^3}. \end{aligned} \tag{3.106}$$

We note that the solution (3.106) may be written as

$$\begin{aligned} v_i &= U\left\{1 - \frac{3a}{4r} - \frac{a^3}{4r^3}\right\}\delta_{i1} + \frac{3}{4}U\frac{ax_1x_i}{r^3}\left(\frac{a^2}{r^2} - 1\right), \\ p &= -\frac{3\mu U a x_1}{2r^3}. \end{aligned} \tag{3.107}$$

To calculate the drag on the sphere, we note, from (3.107), that when $r = a$ we have

$$\begin{aligned} \frac{\partial v_i}{\partial x_j} &= \frac{3}{4}U\left\{1 + \frac{a^2}{r^2}\right\}\frac{ax_j}{r^3}\delta_{i1} + \frac{3}{4}U\frac{ax_1x_i}{r^3}\left(-\frac{2a^2x_j}{r^4}\right) + \text{zero terms}, \\ &= \frac{3}{2}U\frac{x_j\delta_{i1}}{a^2} - \frac{3}{2}U\frac{x_1x_ix_j}{a^4}. \end{aligned}$$

Then the force per unit area exerted on the sphere is given, by $F_i = l_j p_{ij}$, where l_j is the direction cosines, as

$$\begin{aligned} F_i &= l_j p_{ij} \\ &= -\frac{x}{a} p_{ij} \\ &= \frac{1}{a}\left(-x_i p + \frac{3}{2}\mu U \frac{x_j^2 \delta_{i1}}{a^2} + \frac{3}{2}\mu U \frac{x_i x_j \delta_{j1}}{a^2} - 3\mu U \frac{x_1 x_i x_j^2}{a^4}\right) \\ &= \frac{1}{a}\left(-x_i p + \frac{3}{2}\mu U \delta_{i1} - \frac{3}{2}\mu U \frac{x_1 x_i}{a^2}\right), \end{aligned}$$

remembering throughout that all quantities are to be evaluated at $r = a$. Upon substituting for p from (3.107 we find that

$$F_i = \frac{3\mu U}{2a} \delta_{i1},$$

showing that the resultant stress at any point on the body is purely in the x_1 direction, is uniform and is a drag. The total drag experienced by the sphere is then

$$D = \frac{3\mu U}{2a} \times \text{(surface area of sphere)} = 6\pi \mu U a. \tag{3.108}$$

We may alternatively express this as drag coefficient, dividing by $\frac{1}{2}\rho U^2 S$, where S is equal to the area of the maximum section perpendicular to the stream. This yields

$$C_D = \frac{D}{\frac{1}{2}\rho U^2 (\pi a^2)} = \frac{24}{R_a}. \tag{3.109}$$

Equations (3.108) and (3.110) are, like the whole analysis, valid only for very small values of Reynolds number. Comparison with experimental data indicates that agreement is good when

$$R_a < 0.5.$$

3.11.2 Stokes flow using vector calculus

(a) Flow past a sphere

In this section we shall use Stokes's equation for steady flow in the vector form

$$\nabla \cdot \mathbf{v} = 0, \tag{3.110}$$

$$\nabla \times \boldsymbol{\omega} = -\frac{1}{\mu} \nabla p. \tag{3.111}$$

These equations are also valid for unsteady flow if the frequency is such that $\frac{\omega l^2}{\nu} = 0(1)$. It follows immediately from (3.111) that

$$\nabla^2 p = 0, \tag{3.112}$$

$$\text{and} \quad \nabla \times (\nabla \times \boldsymbol{\omega}) = 0. \tag{3.113}$$

Equations (3.112) show that p is a harmonic function, and this result is the starting point for solving Stokes's equations in terms of spherical harmonic function. Referring to steady streaming motions we shall now describe the Stokes flow past a fixed sphere of radius a in terms of the Stokes stream function ψ. In spherical polar coordinates r, θ, ϕ, with the axis $\theta = 0$ chosen to lie in the direction of the free stream U , where $x = r \sin\theta \cos\phi$, $y = r \sin\theta \sin\phi$, and $z = r \cos\theta$, the equation of continuity (3.110) is satisfied if the velocity components are given in terms of ψ by

$$\begin{aligned} v_r &= \frac{1}{r^2 \sin\theta} \frac{\partial \psi}{\partial \theta}, \\ v_\theta &= -\frac{1}{r \sin\theta} \frac{\partial \psi}{\partial r}. \end{aligned} \tag{3.114}$$

In the absence of swirling , $v_\phi = 0$, and the only component of vorticity is

$$\omega_\phi = -\frac{1}{r \sin\theta} \mathcal{D}^2 \psi, \tag{3.115}$$

where $\mathcal{D}^2$ represents the differential operator

$$\mathcal{D}^2 = \frac{\partial^2}{\partial r^2} + \frac{1-\lambda^2}{r^2} \frac{\partial^2}{\partial \lambda^2},$$

in which $\lambda = \cos\theta$. Equation (3.113) reduces to

$$\mathcal{D}^4 \psi = 0, \tag{3.116}$$

and the required solution of this equation satisfying zero velocity at the surface $r = a$, and the velocity $U\mathbf{i}$ at infinity, is

$$\psi = \frac{1}{2} U a^2 \sin^2\theta \left(\frac{r}{a} - 1\right)^2 \left(1 + \frac{a}{2r}\right). \tag{3.117}$$

It follows that

$$\omega_\phi = -\frac{3}{2} U \frac{a}{r^2} \sin\theta, \tag{3.118}$$

and then it is easy to establish from (3.111) that

$$p - p_\infty = -\frac{3}{2} \mu U \frac{a}{r^2} \cos\theta. \tag{3.119}$$

Since ψ is symmetrical above the plane $\theta = \frac{\pi}{2}$, the streamline distribution is symmetrical before and after and therefore shows no wake. This defect in the Stokes solution arises because the convection of vorticity was eliminated from the problem when the inertia term was discarded.

The stream function (3.117) can be written in two parts. The first,

$$\psi = \frac{1}{2} U r^2 \sin^2\theta \left(1 + \frac{1}{2}(a/r)^3\right),$$

represents irrotational flow past a doublet of moment $2\pi U a^3\mathbf{i}$; this makes no contribution to the total force on the sphere. The second part, $\psi = -\frac{3}{4}Uar\sin^2\theta$, represents rotational flow with radial and transverse velocity component equal to $-\frac{3}{2}U(a/r)\cos\theta$ and $\frac{3}{4}U(a/r)\sin\theta$ respectively, and involve a singularity at $r = 0$ that has been called a Stokeslet. The stress on the surface of the sphere has components p_∞ radially inward and $\frac{3}{2}\mu U/a$ downstream, and so the action of the viscosity is to produce a drag

$$D = 6\pi\mu U a. \tag{3.120}$$

This drag is entirely associated with the flow due to the Stokeslet, and it follows that a Stokeslet may be interpreted physically as a force applied to the fluid at a point. For the sphere, the drag coefficient is

$$C_D = \frac{D}{\frac{1}{2}\rho U^2(\pi a^2)} = \frac{24}{R_a}, \tag{3.121}$$

where the Reynolds number R_a is based on diameter of the sphere. The formula (3.121) agrees with the experimental measurement of the drag of sphere for $R_a < 1$, but that it begins to underestimate the drag when $R_a = 1$.

(b) Flow past a circular cylinder

In two-dimensional flow parallel to the x, z plane, the vorticity η is given in terms of the Lagrangian stream function ψ by the relation $\eta = \nabla^2\psi$, and (3.113) becomes

$$\nabla^4\psi = 0. \tag{3.122}$$

For flow past a circular cylinder $r = a$, the stream function representing the free stream at infinity is $Ur\sin\theta$ in the plane polar coordinates r, θ, and so we seek a solution of (3.122) in the form $\psi = f(r)\sin\theta$. It then appears that the general form of $f(r)$ can be written as

$$f(r) = A_1 r^3 + A_2 r\ln r + A_3 r + A_4\frac{1}{r},$$

and the first two terms would have to be omitted, and put $A_3 = U$, in order to match the free stream at infinity. This would leave only A_4 to be determined, and it would be impossible for both v_r and v_θ to vanish at $r = a$. The solution which satisfies the no-slip condition and tends to infinity most slowly as $r \to \infty$ is, in fact,

$$\psi = A\sin\theta\left\{\left(\frac{r}{a}\right)\ln\left(\frac{r}{a}\right) - \frac{1}{2}\left(\frac{r}{a}\right) + \frac{1}{2}\frac{a}{r}\right\}, \tag{3.123}$$

obtained by discarding only the term involving r^3. This expression for ψ leads to a definite formula for the drag, namely $4\pi\mu A/a$, and the expression is valid at points not too far from the cylinder, but of course the solution suffers from the defect that it does not determine the value of the constant A. This solution appears to be the first approximation, and the value of A must be obtained matching with the second approximation. We do not want to pursue the matter any further. But the interested reader is referred to Rosenhead's Boundary Layer Theory [10].

3.11.3 Oseen flow

We have seen in the Stokes flow, there are sever limitations of the solution. These limitations may be recognized again by noticing that $\mathbf{v} \times \boldsymbol{\omega}$, a typical inertial term neglected, has a magnitude $\frac{Ur}{\nu} = Rr/l$ compared with $\nu\nabla \times \boldsymbol{\omega}$, the viscous stress represented in Stokes's equation. However, no matter how small R may be, the assumptions underlying Stokes's equations are not valid at sufficiently large distances r from obstacles. To avoid this difficulty Oseen [15] proposed that the inertial terms should be retained in the far field where the velocity is approximately equal to $U\mathbf{i}$. These inertia terms are of the order R near the obstacle, where it is permissible to neglect them altogether, and so we find that in three-dimensional flow Stokes's and Oseen's equations both yield the same terms of order 1, and only differ in terms of order R.

Oseen's equations for steady flow are given in the following form

$$\nabla \cdot \mathbf{v} = 0, \tag{3.124}$$

$$\begin{aligned} U\left(\frac{\partial \mathbf{v}}{\partial x}\right) &= -\frac{1}{\rho}\nabla p - \nu\nabla \times \boldsymbol{\omega} \\ &= -\frac{1}{\rho}\nabla p + \nu\nabla^2\mathbf{v}. \end{aligned} \tag{3.125}$$

By taking the divergence of (3.125) we find that

$$\nabla^2 p = 0, \tag{3.126}$$

and, by taking the curl, that

$$\left\{\nabla^2 - 2k\frac{\partial}{\partial x}\right\}\boldsymbol{\omega} = 0, \tag{3.127}$$

where $k = \frac{U}{2\nu}$. It follow from (3.126) that p can be expressed by

$$p - p_\infty = -\rho U\left(\frac{\partial \phi}{\partial x} - U\right), \tag{3.128}$$

where ϕ satisfies

$$\nabla^2\phi = 0. \tag{3.129}$$

It then follows that the velocity vector can be assumed as

$$\mathbf{v} = \nabla\phi + \frac{1}{2k}\nabla\chi - \mathbf{i}\chi, \tag{3.130}$$

provided that

$$\left(\nabla^2 - 2k\frac{\partial}{\partial x}\right)\chi = 0. \tag{3.131}$$

The flow field is seen to consist of two components, an irrotational flow with velocity $\mathbf{v}_1 = \nabla\phi$ satisfying

$$\nabla \cdot \mathbf{v}_1 = 0,$$
$$U\left(\frac{\partial \mathbf{v}_1}{\partial x}\right) = -\frac{1}{\rho}\nabla p, \tag{3.132}$$

and a rotational flow with velocity $\mathbf{v}_2 = \frac{1}{2k}\nabla\chi - \mathbf{i}\chi$ satisfying

$$\nabla \cdot \mathbf{v}_2 = 0$$
$$U\left(\frac{\partial \mathbf{v}_2}{\partial x}\right) = -\nu\nabla \times \boldsymbol{\omega} = \nu\nabla^2\mathbf{v}_2. \tag{3.133}$$

Thus the pressure is entirely associated with the irrotational flow, while the vorticity in the rotational flow is given by

$$\boldsymbol{\omega} = -\nabla \times (\chi\mathbf{i}). \tag{3.134}$$

The representation of the flow in the manner of (3.128) and (3.130) is due to Lamb [4]. Details will not be given here, except for the comment that accurate calculation of flow past a sphere in the Oseen approximation require the use of more basic solution that were needed to obtain the Stokes solution. Among the results that may be obtained, we note that the drag coefficient for uniform flow past a sphere is obtained as

$$C_D = \frac{24}{R_a}\left(1 + \frac{3}{16}R_a + O(R_a^2)\right), \tag{3.135}$$

in which the term $\frac{3}{16}R_a$ is Oseen's correction to the Stokes drag coefficient.

Remark

For two-dimensional flow past a cylinder, Oseen's momentum equations and the continuity equation are satisfied by

$$u = \frac{\partial\phi}{\partial x} + \frac{1}{2k}\frac{\partial\chi}{\partial x} - \chi$$
$$v = \frac{\partial\phi}{\partial y} + \frac{1}{2k}\frac{\partial\chi}{\partial y},$$
$$p = -\rho U\frac{\partial\phi}{\partial x};$$

where $k = \frac{U}{2\nu}$ and U is the undisturbed stream velocity parallel to the $x-$ axis, provided

$$\nabla^2\phi = 0,$$
$$\text{and}\quad \left(\nabla^2 - 2k\frac{\partial}{\partial x}\right)\chi = 0.$$

The lifting force per unit length on the cylinder is given by $\rho U K$, where K is the circulation in a very large contour surrounding the cylinder.

For further information about this topic, the reader is referred to the work of Hellwig, G. [3], Lamb, H. [5], Milne-Thomson, L.M. [7], Rahman, M. [8] and Wylie, C.R. & Barrett, L.C. [13] as listed in the reference section.

3.12 Exercises

1. The velocity potential, $\phi(r,\theta)$, for the two-dimensional irrotational flow of an ideal fluid satisfies Laplace's equation. The velocity of the fluid is $\mathbf{V} = \mathbf{grad}\,\phi$, and the radial velocity is $\frac{\partial\phi}{\partial r}$. If the velocity of the fluid at infinity is parallel to $\theta = 0$ and is $\mathbf{V} = Ui$, and if the flow passes around a circular cylinder $r = a$ on which the boundary condition is $\frac{\partial\phi}{\partial r} = 0$, confirm that the velocity potential is $\phi = Urcos\theta + U\frac{a^2}{r}cos\theta$.
2. The velocity potential due to a two-dimensional point source (a line source normal to the x, y plane) of strength m at the point $x = 0, y = d$ in an infinite fluid is $\phi = mln\{x^2 + (y-d)^2\}^{1/2}$. If a fixed boundary, $y = 0$, is inserted into the flow giving rise to the boundary condition $\frac{\partial\phi}{\partial y} = 0$ on $y = 0$, show that the potential in the region $y > 0$ is $\phi = mln\{x^2 + (y-d)^2\}^{1/2} + mln\{x^2 + (y+d)^2\}^{1/2}$.

   Interpret the second term as an image source at $x = 0, y = -d$. Determine the flow in the quadrant $x \geq 0, y \geq 0$, with the boundary conditions $\frac{\partial\phi}{\partial x} = 0$ on $x = 0$, $\frac{\partial\phi}{\partial y} = 0$ on $y = 0$, due to a fluid source of strength m at the point $x = a, y = b$ of the quadrant.
3. The Legendre polynomials, P_0, P_1, P_2, are $P_0(cos\theta) = 1$, $P_1(cos\theta) = cos\theta$, $P_2(cos\theta) = \frac{1}{2}(3cos^2\theta - 1)$. Hence, the simplest axially symmetric solutions of Laplace's equation in spherical polar coordinates are $A_0 + \frac{B_0}{r}$, $\left(A_1 r + \frac{B_1}{r^2}\right)cos\theta, and \left(A_2 r^2 + \frac{B_2}{r^3}\right)(3cos^2\theta - 1)$. The irrotational flow of an ideal fluid has the velocity field $\mathbf{V} = \mathbf{grad}\,\phi$, where ϕ is the velocity potential which satisfies $\nabla^2\phi = 0$. Confirm that the uniform flow field, $\mathbf{V} = U\mathbf{k}$, corresponds to the case $\phi = Urcos\theta$ where U is a constant. Similarly show that if a rigid sphere of radius $r = a$ is placed in this uniform flow field then the potential becomes $\phi = U(r + a^3/2r^2)cos\theta$, where the boundary condition of zero normal flow on the sphere is $\frac{\partial\phi}{\partial r} = 0$ on $r = a$.
4. If $\phi = r^n s$ is a spherical harmonic, prove that, s being independent of r, $\frac{1}{sin\theta}\frac{\partial}{\partial\theta}(sin\theta\frac{\partial s}{\partial\theta}) + \frac{1}{sin^2\theta}\frac{\partial^2 s}{\partial\omega^2} + n(n+1)s = 0$, Deduce that s/r^{n+1} is also a spherical harmonic. (Note: ϕ is defined to be spherical harmonic provided it satisfies Laplace's equation in spherical coordinates).
5. If $\phi = r^n s$ is a spherical harmonic symmetrical about the x-axis, and s is independent of r, show that $\frac{d}{d\mu}[(1-\mu^2)\frac{ds}{d\mu}] + n(n+1)s = 0$ where $\mu = cos\theta$. Show that the solutions of this equation corresponding to $n = 0$ and $n = 1$, are $P_0(\mu)$, $P_1(\mu)$, show that $Q_0(\mu) = \frac{1}{2}ln\frac{1+\mu}{1-\mu}$, $Q_1(\mu) = \frac{1}{2}ln\frac{1+\mu}{1-\mu} - 1$

6. The motion of fluid is given by the velocity potential $\phi = C\{(1+\frac{1}{n})\frac{r^n}{a^{n-1}} + \frac{a^{n+2}}{r^{n+1}}\}P_n(cos\theta)$, in which C is a constant, and r and θ are spherical polar coordinates. Determine the stream function.
7. A sphere of radius a is surrounded by a concentric spherical shell of radius b, and the space between is filled with liquid. If the sphere is moving with velocity V, show that $\phi = \frac{Va^3}{b^3-a^3}\{r + \frac{b^3}{2r^2}\}cos\theta$ and find the current function.
8. A solid sphere of radius a moves in a fluid which far from the sphere remains at rest at pressure P_0. At time the centre of the sphere is at $(csin\sigma t, 0, 0)$. Show that the pressure at the point $\mathbf{r}(x, y, z)$ at time $t = (\frac{\pi}{2\sigma})$, when the sphere is instantaneously at rest, is $p = p_0 - \frac{1}{2}ca^3\sigma^2(x/r^3)$.
9. Find the values of A and B for which $(Ar + \frac{B}{r^2})cos\theta$ is the velocity function of the motion of an incompressible fluid which fills the space between a solid sphere of radius a, and a concentric spherical shell of radius 2a. The sphere has a velocity U and the shell is at rest. Prove that the kinetic energy of the fluid of density ρ is $10\pi\rho a^3U^2/21$.
10. A sphere, of mass M and radius a, is at rest with its centre at a distance h from a plane boundary. Show that the magnitude of the impulse necessary to start the sphere with a velocity V directly towards the boundary is, very nearly, $V\{M + \frac{1}{2}M'(1 + \frac{3a^3}{8h^3})\}$, where M' is the mass of the displaced fluid. Find also the impulse on the plane boundary.
11. A sphere of radius a moves in a semi-infinite liquid of density ρ bounded by a plane wall, its centre being at a great distance h from the wall. Show that the approximate kinetic energy of the fluid is $1/3\pi\rho a^3V^2\{(1 + \frac{3}{16}\frac{a^3}{h^3}(1 + sin^2\alpha)\}$. The sphere is moving at an angle α with the wall, at a speed V.
12. The ellipsoid $x^2/a^2 + y^2/b^2 + z^2/c^2 = 1$ is placed in a uniform stream parallel to the x-axis. Prove that the lines of equal pressure on the ellipsoid are its curves of intersection with the cones $y^2/b^2 + z^2/c^2 = x^2/h^2$, where h is an arbitrary constant.

References

[1] Carslaw, H.S. & Jaeger, J.C., *Conduction of Heat in Solids*, Clarendon Press: Oxford, 1947.

[2] Curle, N. & Davies, H., *Modern Fluid Dynamics*, Vol. 1, D. Van Nostrand: London, 1968.

[3] Hellwig, G., *Partial Differential Equations*, Blaisdell: Waltham, Massachusetts, 1964.

[4] Lamb, H., On the uniform motion of a sphere in a viscous fluid. *Phil. Mag., (6)*, **21**, pp. 112–119, [176, 180], 1911.

[5] Lamb, H., *Hydrodynamics*, 6th edn, Cambridge University Press: Cambridge, 1932.

[6] Love, A.E.H., *Treaties on the Mathematical Theory of Elasticity*, Cambridge University Press: Cambridge, 1927.

[7] Milne-Thomson, L.M., *Theoretical Hydrodynamics*, 5th edn, MacMillan: New York, 1968.

[8] Rahman, M., *Mathematical Methods with Applications*, WIT Press: Southampton, UK, 2001.

[9] Rahman, M., *Water Waves: Relating Modern Theory to Advanced Engineerinr Applications*, Clarendon Press: Oxford, 1995.

[10] Rosenhead, L. (Ed.), *Laminar Boundary Layers*, Clarendon Press: Oxford, 1963.

[11] Stokes, G.G., On the theory of oscillatory waves, *Trans Camb. Phil. Soc.*, **8**, pp. 441–455, 1847.

[12] Taylor, G.I., *The Determination of Stresses by Means of Soap Films, The Mathematical Properties of Fluids (A Collective Work)*, Blackie and Son: London, 1937.

[13] Wylie, C.R. & Barrett, L.C., *Advanced Engineering Mathematics*, McGraw-Hill: New York, 1982.

[14] Rahman, M., *Applied Differential Equations for Scientists and Engineers*, Vols.1 & 2, Computational Mechanics Publications: Southampton, UK, 1991.

[15] Oseen, C. W., Über die Stokesschen Formel und Über eine Verwandte Aufgabe in der Hydrodynamik. *Ark. Mat. Astr. Fys.*, **6(29)**, 1910.

CHAPTER 4

Laminar boundary layers

Pierre Simon Laplace

Pierre Simon Laplace (1749–1827) is well known for the equation that bears his name. The Laplace equation is one of the most ubiquitous equations of mathematical physics; it appears in electrostatics, electro-magnetics, hydrodynamics, groundwater flow, thermodynamics, and many other fields. As had Euler, Laplace worked in a great variety of areas, applying his knowledge of mathematics to physical problems. He has been called the Newton of France. Laplace was born in Beaumont-en-Auge, Normandy, France, and educated at Caen (1765–1768). In 1768 he became Professor of Mathematics at École Militaire in Paris. Later, he moved to École Normale, also in Paris. Napoleon appointed him Minister of Interior in 1799, and he became a Count in 1806 and a Marquis in 1807; in the same year he assumed the presidency of the French Academy of Sciences. Laplace devoted considerable time doing research on astronomy. He wrote on orbital motion of the planets and celestial mechanisms and on the stability of the solar system.

4.1 Introduction

4.1.1 The concept of the boundary layer

In the previous chapter we have discussed flows with slow motion as manifested by Stokes and Oseen. It is now evident that there must always be a region of slow-moving fluid close to a solid body in which the approximation of inviscid flow breaks down. In 1904, Ludwig Prandtl was the first scientist who observed that these regions of slow-moving fluid are frequently thin. Thus, close to the wall of a body in the fluid, the velocity component parallel to the wall rises rapidly from a value 0 at the wall itself to a value U within a short distance, say δ, from the wall. Accordingly, the velocity gradient $\frac{\partial u}{\partial y}$ is large, and the viscous stress $\mu\frac{\partial u}{\partial y}$ becomes important even when μ is small.

It is fairly straightforward to indicate, by dimensional analysis, that the ratio $\frac{\delta}{L}$ of the length scale normal to and parallel to the wall is of the order $R^{-\frac{1}{2}}$, where $R = \frac{UL}{\nu}$ is the Reynolds number. To see this point clearly, let us consider two-dimensional incompressible flow, and look first of all at the equation of continuity $\nabla \cdot \mathbf{v} = 0$, which can be written in two-dimensional form in x, y coordinates with velocity components (u, v) as

$$\frac{\partial u}{\partial x} + \frac{\partial v}{\partial y} = 0 \tag{4.1}$$

Consider that the typical value of u to be U, and the typical length scale parallel to the wall to be L, and it follows that the order of $\frac{\partial u}{\partial x} = O(\frac{U}{L})$. But a typical length scale normal to the wall is δ, and hence (4.1) shows that

$$v = O\left(\frac{U\delta}{L}\right), \tag{4.2}$$

so that the velocity component normal to the wall is small if $\frac{\delta}{L}$ is small. To make the order analysis of the momentum equations, we consider the first momentum equation in two-dimensions which is simply

$$u\frac{\partial u}{\partial x} + v\frac{\partial u}{\partial y} = -\frac{1}{\rho}\frac{\partial p}{\partial x} + \nu\left(\frac{\partial^2 u}{\partial x^2} + \frac{\partial^2 u}{\partial y^2}\right). \tag{4.3}$$

By using the order of magnitude (4.2) for v, and by assuming that a typical value of the pressure p is ρU^2, we can deduce that the magnitudes of the five terms in

(4.3) are of the order

$$u\frac{\partial u}{\partial x} = U\left(\frac{U}{L}\right) = \frac{U^2}{L},$$

$$v\frac{\partial u}{\partial y} = \frac{U\delta}{L}\left(\frac{U}{\delta}\right) = \frac{U^2}{L},$$

$$\frac{1}{\rho}\frac{\partial p}{\partial x} = \frac{U^2}{L},$$

$$\nu\frac{\partial^2 u}{\partial x^2} = \frac{\nu U}{L^2},$$

$$\nu\frac{\partial^2 u}{\partial y^2} = \frac{\nu U}{\delta^2}.$$

Using these orders in (4.3), we see the proportions of the five terms of momentum equation in the following order

$$1 : 1 : 1 : \frac{1}{R} : \frac{1}{R}\left(\frac{L}{\delta}\right)^2.$$

We note that when the Reynolds number R is large, the viscous terms are both small when δ and L have the same order. However, if R is large but

$$\frac{1}{R}\left(\frac{L}{\delta}\right)^2 = O(1),$$

that means
$$\frac{\delta}{L} = O\left(R^{-\frac{1}{2}}\right), \tag{4.4}$$

then although the term $\nu\frac{\partial^2 u}{\partial x^2}$ is small, the term $\nu\frac{\partial^2 u}{\partial y^2}$ is not small. This thin region, in which the viscous effects are important, is called a **boundary layer**.

A more precise criterion for the existence of a well-defined laminar boundary layer is that the Reynolds number should be large, but should not be so large as to breakdown of the laminar layer. This is one of the important features of the exact solutions of Navier-Stokes equations discussed in the last chapter. For the large values of R, these solutions exactly exhibit regions of steep velocity gradient in the boundary layer in the sense already defined. To this theoretical evidence may be added the results of much experimental observation, which shows that these boundary layers exist in practice.

4.1.2 Mathematical expression of the boundary-layer thickness $\delta(x)$

A method of indicating the possibility of a thin boundary-layer of thickness given by (4.4) is as follows. Let us consider the motion of a fluid bounded by a semi-infinite plate, $x > 0$, $y = 0$, the flow far from the plate being uniform, parallel to the plate, and with velocity U. If we assume that the flow starts impulsively from rest at time $t = 0$, initially the flow will be completely irrotational and in fact completely uniform, with a vortex sheet of appropriate strength on the plate, across which the velocity parallel to the plate rises rapidly from zero on the plate itself to the free stream velocity U. We know that the viscosity has the diffusive property just like heat diffusion, this vortex sheet will not remain of zero thickness; due to diffusive nature, its thickness at time $t > 0$ will be of the order $(\nu t)^{\frac{1}{2}}$. As we know that the time needed for fluid to travel a distance x along the plate is of the order $\frac{x}{U}$, and in this time the vortex sheet has grown to a thickness of the order $(\frac{\nu x}{U})^{1/2}$, which is precisely the order of boundary-layer thickness, i.e., $\delta(x) = (\frac{\nu x}{U})^{1/2}$.

4.1.3 Boundary layer separation

There will exist a stage of the fluid flow when a back-flow takes place close to the wall, the forward-moving fluid in the mainstream is forced to move away or separate from the wall in order to by-pass the reverse-flow region. From theoretical point of view we may say that all available calculations relevant to separating laminar boundary layer indicate that the reverse-flow velocities is very low, but persists over a considerable distance from the wall. This is certainly borne out by the available experimental evidence, which indicates that the Reynolds number s, which are in practical situations boundary layer separation usually causes a distortion of the mainstream sufficient to make it differ considerably from that given the inviscid flow past the same body. Accordingly, if boundary layer separation does occur, it may be impossible to calculate the flow pattern on a purely theoretical basis; it certainly is impossible to calculate an inviscid flow and then add the effects of the boundary layer.

To have some more practical insight about the laminar boundary theory, we can consider the flow through a diffuser, that is through a pipe of increasing cross-section. In steady motion, the mass of fluid flowing across any cross-section must remain constant. Thus, if the fluid follows on the walls, as it would do in inviscid flow, the velocity must decrease and the pressure must increase with increasing cross-section. Accordingly, the fluid is moving against the pressure gradient, and the thickness of the boundary layers on the walls may be expected. If the expansion of the tube is rapid and the adverse pressure gradient is correspondingly great, we may expect boundary layer separation.

A further practical manifestation of the boundary layer separation is in the stalling of the flow past a thin aerofoil as the thickness increases. For a carefully designed streamline body such as an aerofoil, that at small angles of incidence, the pressure on the upper surface falls sharply from a maximum at a forward stagnation point and then rises again downstream of a suction peak, the rise being relatively sharp initially and gradually becoming gentler further back. On the lower surface of the aerofoil, the pressure gradient is everywhere favourable downstream of the pressure peak at the forward stagnation point so no tendency to separate exists.

4.2 Derivation of the boundary layer equations for flow along a flat plate

In developing a mathematical theory of boundary layers, the first step is to show the existence, as the Reynolds number R tends to infinity, or the kinematic viscosity ν tends to zero in a limiting form but not just putting $\nu = 0$. A solution of these limiting equations may then reasonably be expected to describe approximately the flow in a laminar boundary layer for which R is large but not infinity.

The full equations of motion for two-dimensional flow are

$$\frac{\partial u}{\partial t} + u\frac{\partial u}{\partial x} + v\frac{\partial u}{\partial y} = -\frac{1}{\rho}\frac{\partial p}{\partial x} + \nu\left(\frac{\partial^2 u}{\partial x^2} + \frac{\partial^2 u}{\partial y^2}\right), \tag{4.5}$$

$$\frac{\partial v}{\partial t} + u\frac{\partial v}{\partial x} + v\frac{\partial v}{\partial y} = -\frac{1}{\rho}\frac{\partial p}{\partial y} + \nu\left(\frac{\partial^2 v}{\partial x^2} + \frac{\partial^2 v}{\partial y^2}\right), \tag{4.6}$$

$$\frac{\partial u}{\partial x} + \frac{\partial v}{\partial y} = 0. \tag{4.7}$$

The $x - y$ plane is taken as the plane of the boundary-layer flow, with the axis of x along, and that of z perpendicular to, the plane wall; thus $u = v = 0$ on $y = 0$.

The equation of conservation of mass (4.7) implies that there exists a stream function $\psi(x, y, t)$ such that

$$u = -\frac{\partial \psi}{\partial y}, \quad v = \frac{\partial \psi}{\partial x}. \tag{4.8}$$

and a fourth-order equation for ψ may be obtained by eliminating the pressure p from equations (4.5) and (4.6), and using (4.8). It is convenient to use the vector notation of the continuity and momentum equations in two-dimensions in the following manner.

$$\nabla \cdot \mathbf{v} = 0, \tag{4.9}$$

$$\frac{\partial \mathbf{v}}{\partial t} + (\mathbf{v}\cdot\nabla)\mathbf{v} = -\frac{1}{\rho}\nabla p + \nu\nabla^2\mathbf{v}. \tag{4.10}$$

The vorticity vector is

$$\boldsymbol{\omega} = \nabla \times \mathbf{v} = -\mathbf{k}\nabla^2\psi.$$

Using vector identities, the second term, i.e., the convective inertia term can be simplified as

$$(\mathbf{v}\cdot\nabla)\mathbf{v} = \nabla\left(\frac{1}{2}q^2\right) - \mathbf{v}\times\boldsymbol{\omega}.$$

Substituting this information in (4.10), and after taking curl of the resulting equation yields

$$\frac{\partial \boldsymbol{\omega}}{\partial t} - \nabla\times(\mathbf{v}\times\boldsymbol{\omega}) = \nu\nabla\times(\nabla^2\mathbf{v}), \tag{4.11}$$

which can be subsequently reduced to the following simple form

$$\frac{\partial \boldsymbol{\omega}}{\partial t} + (\mathbf{v} \cdot \nabla)\boldsymbol{\omega} - (\boldsymbol{\omega} \cdot \nabla)\mathbf{v} = \nu \nabla \times (\nabla^2 \mathbf{v}) = \nu \nabla^2 \boldsymbol{\omega}. \tag{4.12}$$

Since the only component of vorticity is $\eta = \frac{\partial u}{\partial y} - \frac{\partial v}{\partial x} = \nabla^2 \psi$, (4.12) may also be written in the form

$$\frac{\partial \eta}{\partial t} + \frac{\partial \psi}{\partial y}\frac{\partial \eta}{\partial x} - \frac{\partial \psi}{\partial x}\frac{\partial \eta}{\partial y} = \nu \nabla^2 \eta. \tag{4.13}$$

In order to derive the boundary layer equations, a non-dimensional form of equations is convenient. If L is the typical length, U_0 a typical speed and $R = \frac{UL}{\nu}$ the corresponding Reynolds number for the flow as a whole, the following dimensionless variables are defined:

$$X = \frac{x}{L}, \qquad Y = \frac{R^{\frac{1}{2}} y}{L}$$
$$U = \frac{u}{U_0}, \qquad V = \frac{R^{\frac{1}{2}} v}{U_0}$$
$$T = \frac{tU_0}{L}, \qquad P = \frac{p}{\rho U_0^2}. \tag{4.14}$$

Equations (4.5–4.7) become

$$\frac{\partial U}{\partial T} + U\frac{\partial U}{\partial X} + V\frac{\partial U}{\partial Y} = -\frac{\partial P}{\partial X} + \frac{1}{R}\frac{\partial^2 U}{\partial X^2} + \frac{\partial^2 U}{\partial Y^2}, \tag{4.15}$$

$$\frac{1}{R}\left(\frac{\partial V}{\partial T} + U\frac{\partial V}{\partial Y} + V\frac{\partial V}{\partial Y}\right) = -\frac{\partial P}{\partial Y} + \frac{1}{R^2}\frac{\partial^2 V}{\partial X^2} + \frac{1}{R}\frac{\partial^2 V}{\partial Y^2}, \tag{4.16}$$

$$\frac{\partial U}{\partial X} + \frac{\partial V}{\partial Y} = 0, \tag{4.17}$$

in which quantities denoted by capital letters are non-dimensional. On the assumptions that all the derivatives in these equations are of the same order of magnitude for large values of R, their limiting forms are

$$\frac{\partial U}{\partial T} + U\frac{\partial U}{\partial X} + V\frac{\partial U}{\partial Y} = -\frac{\partial P}{\partial X} + \frac{\partial^2 U}{\partial Y^2}, \tag{4.18}$$

$$0 = -\frac{\partial P}{\partial Y} \tag{4.19}$$

$$\frac{\partial U}{\partial X} + \frac{\partial V}{\partial Y} = 0, \tag{4.20}$$

Equations (4.18–4.20) are the boundary layer equations in non-dimensional form. The original method of deriving the boundary-layer equations, due to Prandtl (1904) and Blasius (1908), is based on a consideration of approximate orders of magnitude. It is less precise but may be physically easier to understand. Suppose that L is the typical length, and U is the typical speed, along the boundary layer; and that δ is a typical length, and V is a typical speed, across the layer. In some sense δ is the *thickness* of the boundary layer. Then in (4.7) the terms $\frac{\partial u}{\partial x}$ and $\frac{\partial v}{\partial y}$ are respectively of order $\frac{U}{L}$ and $\frac{V}{\delta}$, Consequently $\frac{V}{U} = O(\frac{\delta}{L})$, and the ratio $\frac{\delta}{L}$ is supposed to be small compared with unity.

Thus with all these observations and also with the order analysis already manifested, the two-dimensional boundary layer equations in dimensional form are given by

$$\frac{\partial u}{\partial t} + u\frac{\partial u}{\partial x} + v\frac{\partial u}{\partial y} = -\frac{1}{\rho}\frac{\partial p}{\partial x} + \nu\left(\frac{\partial^2 u}{\partial y^2}\right), \tag{4.21}$$

$$0 = -\frac{1}{\rho}\frac{\partial p}{\partial y}, \tag{4.22}$$

$$\frac{\partial u}{\partial x} + \frac{\partial v}{\partial y} = 0. \tag{4.23}$$

Remark

If $U(x, t)$ now denotes the main-stream velocity, so that $-\frac{1}{\rho}\frac{\partial p}{\partial x} = \frac{\partial U}{\partial t} + U\frac{\partial U}{\partial x}$, elimination of pressure from (4.21) gives

$$\frac{\partial u}{\partial t} + u\frac{\partial u}{\partial x} + v\frac{\partial u}{\partial y} - \nu\left(\frac{\partial^2 u}{\partial y^2}\right) = f(x, t) = \frac{\partial U}{\partial t} + U\frac{\partial U}{\partial x}, \tag{4.24}$$

This equation was derived by Kármán (1921) by seeking a solution of (4.13) in the form $\psi = \nu^{\frac{1}{2}}\Psi(x, y\nu^{-\frac{1}{2}}, t)$. It may be seen to be of third-order when writing the terms of the stream function, and this is an important mathematical difference from the full equation (4.13), leading to certain anomalies which are discussed in later section.

4.3 Boundary conditions for steady flow

The steady boundary layer equations for u and v are then given by

$$u\frac{\partial u}{\partial x} + v\frac{\partial u}{\partial y} = -\frac{1}{\rho}\frac{\partial p}{\partial x} + \nu\left(\frac{\partial^2 u}{\partial y^2}\right), \tag{4.25}$$

$$p = p(x), \tag{4.26}$$

$$\frac{\partial u}{\partial x} + \frac{\partial v}{\partial y} = 0. \tag{4.27}$$

By virtue of the continuity equation (4.27), we may introduce a stream function ψ such that

$$u = \frac{\partial \psi}{\partial y} \qquad v = -\frac{\partial \psi}{\partial x}. \tag{4.28}$$

Introducing these relations into (4.25) we obtain a third-order partial differential equation in ψ

$$\frac{\partial \psi}{\partial y}\frac{\partial^2 \psi}{\partial x \partial y} - \frac{\partial \psi}{\partial x}\frac{\partial^2 \psi}{\partial y^2} = -\frac{1}{\rho}\left(\frac{dp}{dx}\right) + \nu \frac{\partial^3 \psi}{\partial y^3}. \tag{4.29}$$

The boundary conditions are that

$$u = v = 0 \quad \text{when} \quad y = 0. \tag{4.30}$$

In addition the velocity $u(x, y)$ must join smoothly on to the main stream velocity for some suitable values of y. From physical consideration it is found that the join must take place asymptotically. That means the third boundary condition becomes

$$u \to U(x) \quad \text{as} \quad y \to \infty. \tag{4.31}$$

The relationship between the pressure $p(x)$ and the external velocity $U(x)$ is given either by reference to Bernoulli's equation for inviscid flow or by letting $y \to \infty$ in (4.25). Thus, since $\frac{\partial u}{\partial y} \to 0$, and $\frac{\partial^2 u}{\partial y^2} \to 0$ as $y \to \infty$, we have then

$$U\frac{dU}{dx} = -\frac{1}{\rho}\frac{dp}{dx}$$

or, upon integration

$$p + \frac{1}{2}\rho U^2 = \text{constant} = p_0 + \frac{1}{2}\rho U_0^2.$$

Hence the pressure coefficient C_p is given by

$$C_p = \frac{p - p_0}{\frac{1}{2}\rho U_0^2} = 1 - \frac{U^2}{U_0^2}.$$

4.4 Boundary layer equations for flow along a curved surface

This section is devoted to the boundary layer equation for flow along a curved surface. The surface may be a surfaces of cylinder, sphere or any other regular geometry. We accordingly choose general orthogonal coordinates in two-dimensional motion of a viscous fluid. The equations of motion and the continuity equation in vector notation are written as

$$\frac{\partial \mathbf{v}}{\partial t} + (\mathbf{v} \cdot \nabla)\mathbf{v} = -\frac{1}{\rho}\nabla p + \nu \nabla^2 \mathbf{v}, \tag{4.32}$$

$$\nabla \cdot \mathbf{v} = 0. \tag{4.33}$$

In two-dimensional curvilinear coordinate system these equations take the following form: (see Rahman [12])

$$\frac{\partial v_1}{\partial t} + v_2\omega_3 = -\frac{1}{h_1}\frac{\partial}{\partial x_1}\left(\frac{p}{\rho} + \frac{1}{2}(v_1^2 + v_2^2)\right) - \frac{\nu}{h_2}\frac{\partial \omega_3}{\partial x_2}, \tag{4.34}$$

$$\frac{\partial v_2}{\partial t} - v_1\omega_3 = -\frac{1}{h_2}\frac{\partial}{\partial x_2}\left(\frac{p}{\rho} + \frac{1}{2}(v_1^2 + v_2^2)\right) + \frac{\nu}{h_2}\frac{\partial \omega_3}{\partial x_2}, \tag{4.35}$$

$$\frac{\partial}{\partial x_1}(h_2 v_1) + \frac{\partial}{\partial x_2}(h_1 v_2) = 0, \tag{4.36}$$

where

$$\omega_3 = \frac{1}{h_1 h_2}\left\{\frac{\partial}{\partial x_1}(h_2 v_2) - \frac{\partial}{\partial x_2}(h_1 v_1)\right\} \tag{4.37}$$

is the only component of vorticity present. Other two components are zero in two-dimensional case.

The coordinates x and y are now defined as distances measured along the wall and at right angles to the wall, so that these form a set of orthogonal curvilinear coordinates. The corresponding velocity components are then u and v, respectively. These coordinates and velocity components are made dimensionless as introduced in the previous section. We introduce the non-dimensional variables X, Y, U, V again here for ready reference. When X and Y are taken for the variables x_1, x_2 and then are used in equations (4.34) to (4.37), we obtain

$$h_1 = L\left(1 + \kappa L R^{-\frac{1}{2}} Y\right) = LH, \quad h_2 = LR^{-\frac{1}{2}}, \tag{4.38}$$

where κ is the curvature of the wall, considered positive in the case of Fig. 4.1, in which the centre of curvature is on the side $y < 0$.

$$X = \frac{x}{L}, \qquad Y = \frac{R^{\frac{1}{2}} y}{L}$$

$$U = \frac{u}{U_0}, \qquad V = \frac{R^{\frac{1}{2}} v}{U_0}$$

$$T = \frac{tU_0}{L}, \qquad P = \frac{p}{\rho U_0^2}.$$

In terms of these non-dimensional variables, (4.34) to (4.36) become

$$H\frac{\partial U}{\partial T} + U\frac{\partial U}{\partial X} + V\frac{\partial (HU)}{\partial Y} = -\frac{\partial P}{\partial X} + H\frac{\partial}{\partial Y}\left\{\frac{1}{H}\frac{\partial (HU)}{\partial Y} - \frac{1}{R}\frac{1}{H}\frac{\partial V}{\partial X}\right\}, \tag{4.39}$$

$$\frac{1}{R}\left\{\frac{\partial V}{\partial T} + \frac{U}{H}\frac{\partial V}{\partial X} + V\frac{\partial V}{\partial Y}\right\} - \frac{U^2}{H}\frac{\partial H}{\partial Y}$$
$$= -\frac{\partial P}{\partial Y} + \frac{1}{H}\frac{\partial}{\partial X}\left\{\frac{1}{R^2}\frac{1}{H}\frac{\partial V}{\partial X} - \frac{1}{R}\frac{1}{H}\frac{\partial (HU)}{\partial Y}\right\}, \tag{4.40}$$

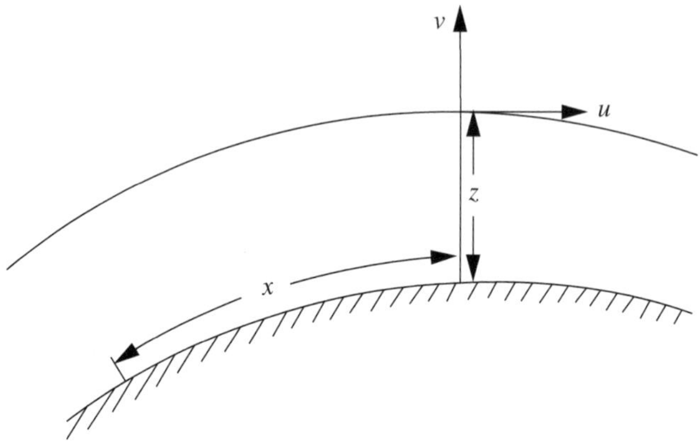

Figure 4.1: Coordinates for boundary-layer flow along a curved surface.

$$\frac{\partial U}{\partial X} + \frac{\partial (HV)}{\partial Y} = 0. \tag{4.41}$$

Now from (4.38), as $\nu \to 0$, so that $R \to \infty$, H tends to 1 and $\frac{\partial H}{\partial X}$, $\frac{\partial H}{\partial Y}$ both tend to 0. Then the limiting forms of (4.39) to (4.41) as $\nu \to 0$ may be seen to be identical with the equations (4.21) to (4.23) obtained in the case of flow over a plane surface. Thus when R is large the equations of motion reduce to the boundary layer equations

$$\frac{\partial u}{\partial t} + u\frac{\partial u}{\partial x} + v\frac{\partial u}{\partial y} = \frac{\partial U}{\partial t} + U\frac{\partial U}{\partial x} + \nu\frac{\partial^2 u}{\partial y^2}, \tag{4.42}$$

$$\frac{\partial u}{\partial x} + \frac{\partial v}{\partial y} = 0, \tag{4.43}$$

which have the same form for flow over a curved surface as for flow past a plane wall. It is, however necessary that $\frac{\partial H}{\partial X}$ and $\frac{\partial H}{\partial Y}$ shall be small compared with 1, so that $\kappa\delta$ and $\delta L d\kappa/dx$ must be small, where δ is a measure of boundary-layer thickness. The pressure gradient across the layer is given by (4.39) and (4.40) as approximately

$$\frac{\partial P}{\partial Y} = \kappa L R^{-\frac{1}{2}} U^2$$

$$\text{or} \quad \frac{\partial p}{\partial y} = \kappa \rho u^2. \tag{4.44}$$

This is just the gradient of pressure required to balance the centrifugal effect of the flow round the curved surface. When dealing with the boundary-layer on a surface whose radius of curvature is continuous and large compared with the boundary-layer thickness, we may treat the problem as being that for a plane surface, although the external flow, and hence the pressure distribution, will depend crucially upon the details of the surface curvature.

4.5 Boundary-layer thicknesses, skin friction, and energy dissipation

In the previous section we have introduced the boundary-layer thickness δ. In most physical problems the solution of the equations (4.42) and (4.43) are such that the velocity component u attains its main-stream velocity U only asymptotically as $R^{\frac{1}{2}}y/L \to \infty$. The thickness of the layer is therefore a nebulous quantity, and is indefinite, as there is always some departure from the asymptotic value at any finite distance y from the surface. It is clear, however, that for all practical purposes the edge of the boundary layer may be defined as being where the velocity has reached 95 per cent of its mainstream value, or perhaps 98 per cent or 99 per cent; in this sense the value of δ is somewhat arbitrary. It would therefore be possible to regard the boundary-layer thickness as the distance δ from the surface beyond which $\frac{u}{U} > 0.99$, for example, but this is not sufficiently precise since $\frac{\partial u}{\partial y}$ is very small there for experimental work, and is not of theoretical significance.

The scale of the boundary-layer thickness can, however, be specified adequately by certain length capable of precise definition, both for experimental measurement and for theoretical study. These measures of boundary-layer thickness are defined as follows:

Displacement boundary-layer thickness δ_1:

$$\delta_1 = \int_0^\infty \left(1 - \frac{u}{U}\right) dy. \tag{4.45}$$

Momentum boundary-layer thickness δ_2:

$$\delta_2 = \int_0^\infty \frac{u}{U}\left(1 - \frac{u}{U}\right) dy. \tag{4.46}$$

Energy boundary-layer thickness δ_3:

$$\delta_3 = \int_0^\infty \frac{u}{U}\left(1 - \frac{u^2}{U^2}\right) dy. \tag{4.47}$$

The upper limit of the integration is taken as infinity to reach the asymptotic value of $\frac{u}{U}$ to 1, but in practice the upper limit is the point beyond which the integrand is negligible.

We can forward some further explanation of how we have arrived at those integral equations. Suppose we consider a particular streamline which is at a distance $h(x, \psi_0)$ from the wall. In inviscid flow, the streamline would have been at a distance $h_i(x, \psi_0)$ from the wall. Thus, by definition, the total mass of fluid flowing in unit time between $y = 0$ and $y = h$ is equal to the mass which would flow between

$y = 0$ and $y = h_i$ inviscid flow with $u = U(x)$ for all y. Thus

$$\int_0^h \rho u dy = \int_0^{h_i} \rho U dy$$

$$\text{or} \quad \int_0^h u dy = U h_i.$$

Hence the amount by which the streamline is displaced outwards under the influence of viscosity is

$$\begin{aligned} h - h_i &= h - \int_0^h \frac{u}{U} dy \\ &= \int_0^h \left(1 - \frac{u}{U}\right) dy \end{aligned}$$

It follows that the amount by which streamlines far from the wall are displaced is

$$\begin{aligned} \lim_{h \to \infty} (h - h_i) &= \delta_1 \\ &= \int_0^\infty \left(1 - \frac{u}{U}\right) dy, \end{aligned}$$

where $\delta_1(x)$ is referred to as the displacement thickness. The significance of this parameter is that it indicates the extent to which the boundary-layer displaces the external flow.

A second important measure of the boundary-layer thickness which has been already defined is the momentum thickness δ_2 as defined in (4.46). We note that

$$\rho U^2 \delta_2 = \int_0^\infty u(\rho U - \rho u) dy,$$

so that $\rho U^2 \delta_2$ is equal to the flux of the defect of momentum in the boundary-layer.

A third important physical measure of boundary-layer thickness is the kinetic-energy thickness defined in (4.47), which measures the flux of the kinetic energy defect within the boundary layer as compared with an inviscid flow.

Two other quantities related to these boundary-layer thicknesses are the *skin friction* τ_w and the *dissipation integral* D. The skin friction is defined as the shearing stress exerted by the fluid on the surface over which it flows, and is therefore the value of p_{yx} at $y = 0$ which is given by

$$\tau_w = \mu \left(\frac{\partial u}{\partial y}\right)_{y=0} \tag{4.48}$$

We consider the following mathematical development to establish the expression for skin friction cited in (4.48). We know that the components of stress in two-dimension are given by

$$p_{11} = p - 2\mu \frac{\partial u}{\partial x},$$

$$p_{12} = p_{21} = -\mu \left(\frac{\partial u}{\partial y} + \frac{\partial v}{\partial x} \right),$$

$$p_{22} = p - 2\mu \frac{\partial u}{\partial y}.$$

Within the boundary layer, $\frac{\partial u}{\partial y}$ is of order U/δ, and $\frac{\partial v}{\partial x}$ is of order $\frac{\delta U}{L^2}$, so that the ratio of these terms is $1 : (\frac{\delta}{L})^2$, that is $1 : R^{-1}$; $\frac{\partial v}{\partial x}$ may therefore be neglected by comparison with $\frac{\partial u}{\partial y}$. Thus using also the two-dimensional continuity equation, we may write

$$p_{11} = p - 2\mu \left(\frac{\partial u}{\partial x} \right)$$

$$p_{12} = p_{21} = -\mu \left(\frac{\partial u}{\partial y} \right)$$

$$p_{22} = p + 2\mu \left(\frac{\partial u}{\partial x} \right).$$

At the wall itself, the stress acting on the wall in the x-direction is simply $-p_{12}$, so we have

$$\tau_w = \mu \left(\frac{\partial u}{\partial y} \right),$$

where τ_w is usually known as the skin-friction or wall shearing stress.

The rate of energy is dissipated by the action of viscosity, which has been found to be $\mu (\frac{\partial u}{\partial y})^2$ per unit time per unit volume,

$$\Phi = \int_0^\infty \mu \left(\frac{\partial u}{\partial y} \right)^2 dy. \tag{4.49}$$

Consequently Φ is the total dissipation in a cylinder of small cross-section with axis normal to the layer per unit time per unit area of cross-section.

4.6 Momentum and energy equations

The skin friction and viscous dissipation are both connected with the boundary layer thicknesses by two equations, which represent the balance of momentum and of energy within a small section of boundary layer. We shall present here the integral equations of these two important quantities.

4.6.1 Momentum integral

The integral is most simply obtained by integration of the boundary layer equations (4.42) and (4.43) as demonstrated by Pohlhausen in 1921. They may be written as

$$-\nu\frac{\partial^2 u}{\partial y^2} = \frac{\partial}{\partial t}(U-u) + U\frac{\partial U}{\partial x} - u\frac{\partial u}{\partial x} - v\frac{\partial u}{\partial y},$$

$$0 = (U-u)\frac{\partial u}{\partial x} + (U-u)\frac{\partial v}{\partial y},$$

By addition of these two equations we obtain

$$-\nu\frac{\partial^2 u}{\partial y^2} = \frac{\partial}{\partial t}(U-u) + \frac{\partial}{\partial x}(Uu-u^2) + (U-u)\frac{\partial U}{\partial x} + \frac{\partial}{\partial y}(vU-uv).$$

On integration with respect to y from 0 to ∞ yields, since $\frac{\partial u}{\partial y}$ and $v(U-u)$ tend to 0 as $y \to \infty$,

$$\nu\left(\frac{\partial u}{\partial y}\right) = \frac{\partial}{\partial t}\int_0^\infty (U-u)dy + \frac{\partial}{\partial x}\int_0^\infty (Uu-u^2)dy + \frac{\partial U}{\partial x}\int_0^\infty (U-u)dy + v_s U, \quad (4.50)$$

where $v_s = v_s(x) = -v(x,0)$ is the velocity of suction. When the integrals are expressed in terms of the boundary layer thicknesses, this becomes

$$\frac{\tau_w}{\rho} = \frac{\partial}{\partial t}(U\delta_1) + \frac{\partial}{\partial x}(U^2\delta_2) + U\frac{\partial U}{\partial x}\delta_1 + v_s U, \quad (4.51)$$

or, in non-dimensional form,

$$\frac{\tau_w}{\rho U^2} = \frac{1}{U^2}\frac{\partial}{\partial t}(U\delta_1) + \frac{\partial \delta_2}{\partial x} + \frac{\delta_1 + 2\delta_2}{U}\frac{\partial U}{\partial x} + \frac{v_s}{U}. \quad (4.52)$$

This equation is referred to as the momentum integral equation and originally derived by Kármán in 1921. The physical interpretation of the equation (4.50) multiplied by $\rho\delta x$ is the equation for the rate of change of momentum defect, $\rho(U-u)$ per unit volume, for the small slice of the boundary layer between the planes x and $x+\delta x$. On the right-hand side of (4.50) the first term represents the local rate of change of the momentum defect, the second term is the rate of change due to convection across the planes x and $x+\delta x$, and the last term represents convection across the porous surface $y=0$. Since there is no convection at the edge of the layer, where the momentum defect is zero, the total rate of change is equal to the opposing frictional force, which gives the term on the left-hand side, together with two terms which cancel outside the boundary layer. One of these is the opposing pressure force

$$\frac{\partial p}{\partial x} = -\rho\left(\frac{\partial U}{\partial t} + U\frac{\partial U}{\partial x}\right)$$

per unit volume, and the other is the rate of change of the main stream momentum ρU associated with any particle of fluid in the boundary layer, namely $\rho(\frac{\partial U}{\partial t} + U\frac{\partial U}{\partial x})$. The net effect of these two is a term

$$-\frac{\partial U}{\partial x}\int_0^\infty (U-u)dy$$

on the left-hand side of (4.50), which appears as the third term on the right-hand side with the opposite sign.

4.6.2 Energy integral

To derive the energy integral, we consider again the equations (4.42) and (4.43). By multiplying these two equations by $2u$ and $(U^2 - u^2)$ respectively, we obtain,

$$-2\nu u\frac{\partial^2 u}{\partial y^2} = 2u\frac{\partial}{\partial t}(U-u) + 2uU\frac{\partial U}{\partial x} - 2u^2\frac{\partial u}{\partial x} - 2uv\frac{\partial u}{\partial y},$$

$$0 = (U^2-u^2)\frac{\partial u}{\partial x} + (U^2-u^2)\frac{\partial v}{\partial y},$$

and by adding they imply

$$2\nu\left(\frac{\partial u}{\partial y}\right)^2 - 2\nu\frac{\partial}{\partial y}\left(u\frac{\partial u}{\partial y}\right) = \frac{\partial}{\partial t}(Uu-u^2) + U^2\frac{\partial}{\partial t}\left(1-\frac{U}{u}\right) + \frac{\partial}{\partial x}(U^2u-u^3) + \frac{\partial}{\partial y}(vU^2 - vu^2)$$

It is to be noted here that since $v(U^2 - u^2)$ and $\frac{\partial u}{\partial y}$ both tend to 0 as $y \to \infty$, this equation gives on integration with respect to y from 0 to ∞,

$$\frac{2\Phi}{\rho} = \frac{\partial}{\partial t}(U^2\delta_2) + U^2\frac{\partial\delta_1}{\partial t} + \frac{\partial}{\partial x}(U^3\delta_3) + v_sU^2. \tag{4.53}$$

Equation (4.53) may be expressed in non-dimensional form as

$$\frac{2\Phi}{\rho U^3} = \frac{1}{U}\frac{\partial}{\partial t}(\delta_1+\delta_2) + \frac{2\delta_2}{U^2}\frac{\partial U}{\partial t} + \frac{1}{U^3}\frac{\partial}{\partial x}(U^3\delta_3) + \frac{v_s}{U}. \tag{4.54}$$

The energy integral may also be regarded as an equation for the 'kinetic energy defect' $\frac{1}{2}\rho(U^2 - u^2)$ per unit volume, namely

$$\frac{\partial}{\partial t}\int_0^\infty \frac{1}{2}\rho(U^2 - u^2)dy + \frac{\partial}{\partial x}\int_0^\infty \frac{1}{2}\rho(U^2 - u^2)udy + \frac{1}{2}\rho U^2 v_s = \Phi + \rho\frac{\partial U}{\partial t}\int_0^\infty (U - u)dy. \quad (4.55)$$

In this equation, the left-hand side represents the sum of the local and convective rates of change of kinetic energy defect. On the right-hand side the first term is the contribution of viscous dissipation. The second is the sum of (a) the rate of energy loss due to pressure gradient forces, which is

$$u\frac{\partial p}{\partial x} = -\rho u\left(\frac{\partial U}{\partial t} + U\frac{\partial U}{\partial x}\right)$$

per unit volume, and (b) the rate of change of main-stream kinetic energy $\frac{1}{2}\rho U^2$ associated with a particle of fluid in the boundary layer, namely

$$\frac{\partial\left(\frac{1}{2}\rho U^2\right)}{\partial t} + u\frac{\partial\left(\frac{1}{2}\rho U^2\right)}{\partial x}$$

per unit volume. As before, these balance outside the boundary layer, but in it have the net effect given by the last term in equation (4.55).

4.7 The von Mises transformation for steady flow

The boundary layer equations in either two-dimensions or in three-dimensions are inherently non-linear. In view of the immense complexity of these boundary layer equations, to solve even the steady two-dimensional incompressible boundary layer equations, various attempts have been made in the past. One of the most useful is that due to von Mises in which new independent variables are considered using the continuity equation.

We know the continuity equation is

$$\nabla\cdot(\rho\mathbf{v}) = \rho\left(\frac{\partial u}{\partial x} + \frac{\partial v}{\partial y}\right) = 0,$$

which is satisfied by the stream function ψ

$$\rho u = \frac{\partial \psi}{\partial y}, \quad \rho v = -\frac{\partial \psi}{\partial x}.$$

We introduce a new independent variable

$$\psi = \int_0^y (\rho u)dy,$$

so that $u(x, y) = u(x, \psi)$ and $\psi = \psi(x, y)$. Thus, $(x, y) \to (x, \psi)$. With this new set of independent variables, the boundary layer equation

$$u\frac{\partial u}{\partial x} + v\frac{\partial u}{\partial y} = U\frac{dU}{dx} + \nu\frac{\partial^2 u}{\partial y^2}$$

can be transformed evaluating the partial derivatives in the equation above.

Thus we calculate according to partial derivative rules,

$$\left(\frac{\partial u}{\partial x}\right)_y = \left(\frac{\partial u}{\partial x}\right)_\psi + \left(\frac{\partial u}{\partial \psi}\right)_x \left(\frac{\partial \psi}{\partial x}\right)_y = \frac{\partial u}{\partial x} - \rho v\frac{\partial u}{\partial \psi}$$

$$\left(\frac{\partial u}{\partial y}\right)_x = \left(\frac{\partial u}{\partial \psi}\right)_x \left(\frac{\partial \psi}{\partial y}\right)_x = \rho u\frac{\partial u}{\partial \psi}$$

Thus, substituting these in the momentum equation above we obtain which simplifies to yield

$$u\frac{\partial u}{\partial x} - U\frac{dU}{dx} = \mu\rho u\frac{\partial}{\partial \psi}\left\{u\frac{\partial u}{\partial \psi}\right\} \tag{4.56}$$

Here the partial derivatives with respect to x now mean derivatives with ψ held constant. This equation is to be solved subject to the boundary conditions

$$u = 0 \text{ when } \psi = 0,$$
$$u = U \text{ as } \psi \to \infty, \tag{4.57}$$

and when u has been determined, the value of v follows from the continuity equation. The von Mises transformation has been used in a number of practical problems specially in respect of heat transfer through a laminar boundary layer.

4.8 Analytical solutions of boundary layer equations

This section is devoted to the analytic solutions of practical problems, which govern the boundary-layer equations. In solving the two-dimensional boundary layer equation, we have to make crucial assumption that the pressure distribution around the body or equivalently the external velocity $U(x)$ is known a priori. Given $U(x)$ then, it is required to solve the boundary-layer equations such as the momentum and continuity equations is subject to the appropriate boundary conditions. One of the most important elements in the development of the methods which are currently used in the boundary layer analysis is to assume the specific convenient forms of the external velocity.

Mathematically speaking, these special solutions are of two types. For example, where the geometrical configuration is extremely simple, such as the problem of flow past a flat plate held parallel to the stream, or where the domain considered is very limited, for example, flow sufficiently near to the forward stagnation point of a

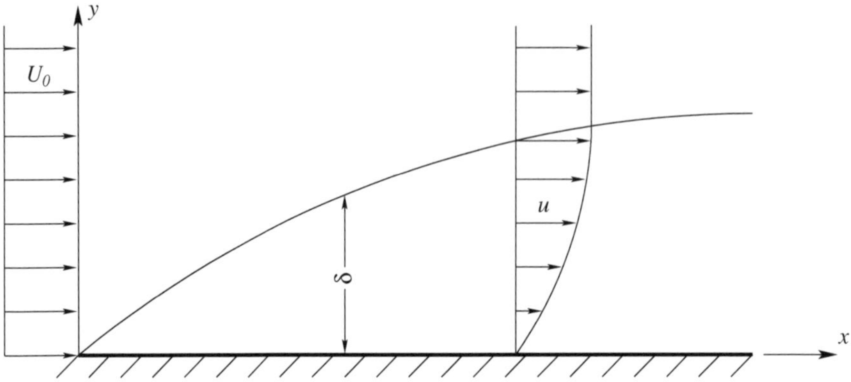

Figure 4.2: The schematic description of laminar boundary layer over a flat plate.

bluff body where the body may in effect be regarded as a plane surface normal to the stream. In those cases, we should look for appropriate transformations to reduce the boundary layer equations to a single ordinary differential equation, whose solution may be obtained to any required accuracy.

The second type of special solution arises when the shape of the body is assumed such that the external velocity $U(x)$ may be expressed as a power series in x containing, say, two or three terms. In those cases, the velocity u in the boundary layer may be expanded as a power series in x, the coefficients being functions of $\frac{y}{\delta}$, say, where δ is representative of the scale normal to the wall.

4.8.1 Flow along a flat plate at zero incidence in a uniform stream

We consider, first of all, what is geometrically the simplest possible configuration, in which a semi-infinite flat plate of zero thickness is placed in a uniform stream U_0. The inviscid-flow solution is, of course, trivial, in that the stream is not affected at all by the plate. The fluid is supposed unlimited, in the external, and the origin of the coordinates is taken at the leading edge, with x measured downstream along the plate and y is perpendicular to it, as shown in Fig. 4.2. This problem was originally solved by Blasius in 1908, and so it is known as the Blasius problem.

In the absence of a pressure gradient, the equations of steady motion in the boundary layer reduces to extremely simple forms

$$u\frac{\partial u}{\partial x} + v\frac{\partial u}{\partial y} = \nu\frac{\partial^2 u}{\partial y^2}, \tag{4.58}$$

$$u = \frac{\partial \psi}{\partial y}, \quad v = -\frac{\partial \psi}{\partial x}, \tag{4.59}$$

and the appropriate boundary conditions are

$$\begin{aligned} y = 0,\ x > 0: &\quad u = v = 0. \\ y = \infty,\ \text{all}\ x: &\quad u = U_0. \\ x = 0: &\quad u = U_0. \end{aligned}$$

These boundary conditions, in fact, demand an infinite gradient in speed at the leading edge $x = y = 0$ implies a singularity in mathematical solution there. However, the assumptions implicit in the boundary layer approximation breaks down for the region of slow flow around the leading edge, and the solution to be derived here must be considered to apply only from a short distance downstream of $x = 0$.

In this type of problem, as found in many mathematical physics problem, we look first for *similarity solutions*, depending not upon x and y independently but only on some combination of x and y. Since the boundary layer velocity profile vary only in scale and is fixed in shape in such a solution, the obvious choice of parameter is $\frac{y}{\delta}$, where δ is representative of the boundary layer thickness. We have already seen that the order of boundary layer thickness is $(\frac{\nu x}{U_0})^{\frac{1}{2}}$ and indeed, for a semi-infinite plate it is impossible to conceive of any other representative length. So the appropriate non-dimensional parameter is

$$y\left(\frac{U_0}{\nu x}\right)^{\frac{1}{2}}.$$

Blasius's solution for the stream function is of the form

$$\psi = (2\nu U_0 x)^{\frac{1}{2}} f(\eta) \tag{4.60}$$

$$\text{where}\quad \eta = \left(\frac{U_0}{2\nu x}\right)^{\frac{1}{2}} y \tag{4.61}$$

although each of these expressions differ by a constant factor $\sqrt{2}$ from Blasius's original definition. Introduction of this factor smooth out the mathematical calculations considerably without any change in the final result. By differentiation, the velocity components are obtained as follows:

$$u = \frac{\partial \psi}{\partial y} = \left(\frac{U_0}{2\nu x}\right)^{\frac{1}{2}} \frac{\partial \psi}{\partial \eta} = U_0 f'(\eta). \tag{4.62}$$

It will be apparent that the particular forms, (4.60) and (4.61), for ψ and η was carefully chosen to yield this simple expression for u, such a form being required, by virtue of the boundary condition that $\frac{u}{U_0} \to 1$ as $y \to \infty$. We now deduce successively that

$$-v = \frac{\partial \psi}{\partial x} = \left(\frac{U_0 \nu}{2x}\right)^{\frac{1}{2}} f - (2U_0 \nu x)^{\frac{1}{2}} \left(\frac{U_0}{2\nu x^3}\right)^{\frac{1}{2}} \frac{1}{2} y f'$$

$$= \left(\frac{U_0 \nu}{2x}\right)^{\frac{1}{2}} (f - \eta f'),$$

$$\frac{\partial u}{\partial x} = -\frac{1}{2} U_0 \left(\frac{U_0}{2\nu x^3}\right)^{\frac{1}{2}} y f'' = -\frac{U_0}{2x} \eta f''$$

$$\frac{\partial u}{\partial y} = U_0 \left(\frac{U_0}{2\nu x}\right)^{\frac{1}{2}} f'',$$

and

$$\frac{\partial^2 u}{\partial y^2} = U_0 \left(\frac{U_0}{2\nu x}\right) f'''.$$

Then upon substitution into (4.58) it is found that the factors which explicitly involve U_0, ν, and x cancel identically, and we obtain

$$f''' + f f'' = 0. \tag{4.63}$$

The boundary conditions become $f(0) = f'(0) = 0$; $f'(\infty) = 1$. Equation (4.63) is usually referred to as the Blasius equation. This is a third-order non-linear ordinary differential equation and there are many ways to solve this equation with the given boundary conditions. The shear stress at any point on the plate is given by

$$\tau = \mu \left(\frac{\partial u}{\partial y}\right)_w = \mu \left(\frac{U_0^3}{2\nu x}\right)^{\frac{1}{2}} f''(0). \tag{4.64}$$

4.8.2 Method of solution

The method of solution of (4.63) given by Blasius required a series expansion for small η which was matched smoothly at a point on to an asymptotic expansion for large η. But the accuracy of the solution depends on the number of terms in each series. A somewhat better method has been used to obtain very accurate solutions.

We first expand in a power series about the origin. We note that $f(0) = f'(0) = 0$ and, and by letting $\eta \to 0$ in (4.63), we further deduce that $f'''(0) = 0$. By differentiating (4.63) successively with respect to η and then letting $\eta \to 0$ we find in turn that $f^{iv}(0) = 0$, $f^{v}(0) = -\left(f''(0)\right)^2$, and also $f^{vi}(0) = f^{vii}(0) = 0$ and so on. By this procedure it is found that

$$f(\eta) = a_0 \alpha \eta^2 + a_1 \alpha^2 \eta^5 + a_2 \alpha \eta^8 + \cdots$$

$$= \sum_0^{\infty} a_n \alpha^{n+1} \eta^{3n+2}, \tag{4.65}$$

where the $a_n's$ are absolute constants, which in fact take values

$$a_0 = \frac{1}{2!},\ a_1 = -\frac{1}{5!},\ a_2 = \frac{11}{8!},\ a_3 = -\frac{375}{11!},$$

and the constant $\alpha = f''(0)$ is determined by the boundary condition at infinity. Now (4.65) may be written as

$$\begin{aligned} f &= \alpha^{\frac{1}{3}} \sum_0^\infty a_n (\alpha^{\frac{1}{3}}\eta)^{3n+2} \\ &= \alpha^{\frac{1}{3}} F(\alpha^{\frac{1}{3}}\eta), \end{aligned} \tag{4.66}$$

where $F(\eta)$ is the solution of equation (4.63) satisfying the boundary conditions $F(0) = F'(0) = 0,\quad F''(0) = 1$. This solution is now easily obtained by step-by-step integration, using the above series as a starting point, and the value of $F'(\infty)$ readily follows. The solution for (4.63) satisfying $f'(\infty) = 1$, is then such that

$$1 = \lim_{\eta\to\infty} (f'(\eta)) = \alpha^{\frac{2}{3}} \lim_{\eta\to\infty} (F'(\alpha^{\frac{1}{3}}\eta) = \alpha^{\frac{2}{3}} F'(\infty),$$

and hence

$$\alpha = (F'(\infty))^{-\frac{3}{2}}.$$

The value of $\alpha = f''(0)$ being now known, it is easy to derive $f(\eta)$ from $F(\eta)$ by using (4.66) or by direct integration of (4.63) step by step. The correct value is

$$\alpha = 0.46960,$$

and the velocity distribution

$$\frac{u}{U_0} = f'(\eta)$$

is shown in Fig. 4.2.

The skin friction is easily calculated from our knowledge of the velocity profiles. Thus we have

$$\begin{aligned} \tau = \mu \left(\frac{\partial u}{\partial y}\right)_w &= \mu \left(\frac{U_0^3}{2\nu x}\right)^{\frac{1}{2}} f''(0) \\ &= 0.33206\mu \left(\frac{U_0^3}{\nu x}\right)^{\frac{1}{2}}. \end{aligned} \tag{4.67}$$

In the non-dimensional form, the skin friction can be expressed as

$$\frac{\tau_w}{\rho U_0^2} = 0.33206 \left(\frac{\nu}{x U_0}\right)^{\frac{1}{2}}.$$

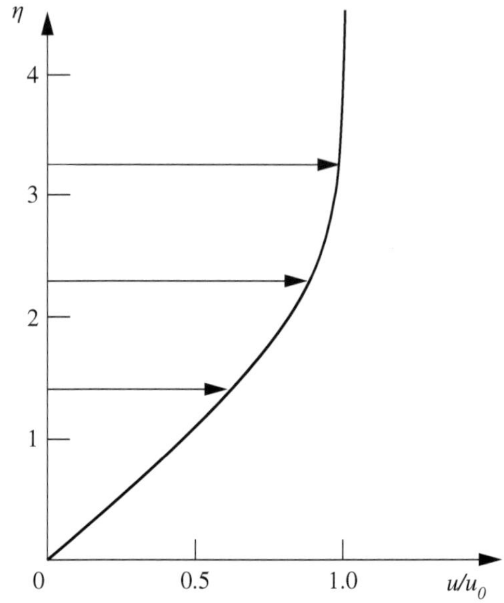

Figure 4.3: The Blasius velocity profile (from [1]).

The drag on one side of a plate of length ℓ and unit breadth is then given by

$$\begin{aligned} D &= \int_0^\ell \tau_w dx \\ &= \mu \left(\frac{U_0^3}{2\nu} \right)^{\frac{1}{2}} f''(0) \int_0^\ell x^{-\frac{1}{2}} dx \\ &= \sqrt{2}(\rho U_0^2 \ell). \left(\frac{U_0 \ell}{\nu} \right)^{-\frac{1}{2}} f''(0), \end{aligned}$$

and the drag coefficient is

$$C_D = \left\{ \frac{D}{\frac{1}{2}\rho U_0^2 \ell} \right\} = 2^{\frac{3}{2}} f''(0) R^{-\frac{1}{2}} = 1.3282 R^{-\frac{1}{2}} \tag{4.68}$$

where $R = \frac{\ell U_0}{\nu}$ = Reynolds number. It should be noted here that the solution is used from $x = 0$, although a slight error may thereby be introduced.

The displacement thickness is, without any approximation,

$$\begin{aligned}\delta_1 &= \int_0^\infty \left(1 - \frac{u}{U_0}\right) dy \\ &= \left(\frac{2\nu x}{U_0}\right)^{\frac{1}{2}} \int_0^\infty (1 - f')d\eta \\ &= \left(\frac{2\nu x}{U_0}\right)^{\frac{1}{2}} . \lim_{\eta\to\infty} (\eta - f) \\ &= 1.72077 \left(\frac{\nu x}{U_0}\right)^{\frac{1}{2}} . \qquad (4.69)\end{aligned}$$

Remark

The numerical values of the following integrals are cited from Rosenhead ([7], p. 224).

$$\int_0^\infty (1 - f')d\eta = \lim_{\eta\to\infty} (\eta - f) = 1.21678$$

$$\int_0^\infty f'(1 - f')d\eta = f''(0) = 0.46960$$

$$\int_0^\infty f'(1 - f'^2)d\eta = 0.73849$$

The momentum thickness is

$$\begin{aligned}\delta_2 &= \int_0^\infty \left(\frac{u}{U_0} - \left(\frac{u}{U_0}\right)^2\right) dy \\ &= \left(\frac{2\nu x}{U_0}\right)^{\frac{1}{2}} \int_0^\infty (f' - f'^2)d\eta \\ &= \left(\frac{2\nu x}{U_0}\right)^{\frac{1}{2}} f''(0) \\ &= 0.6641 \left(\frac{\nu x}{U_0}\right)^{\frac{1}{2}} \qquad (4.70)\end{aligned}$$

Note that in the Blassius equation $f''' + ff'' = 0$ if we add and subtract f'^2 on the left-hand side and integrate from 0 to ∞ we obtain that $\int_0^\infty (f' - f'^2)d\eta = f''(0)$.

An asymptotic form of solution

An asymptotic formula valid for large values of η may be found as follows: in the Blassius equation (4.63) namely $f''' + ff'' = 0$, substitute $f(\eta) = \eta - \beta + \phi(\eta)$, where $\phi(\eta)$ is small. If the product $\phi\phi''$ is neglected, and approximate linear equation for $\phi(\eta)$ is obtained:

$$\phi''' + (\eta - \beta)\phi'' = 0. \tag{4.71}$$

Hence, writing $\xi = \eta - \beta$, it follows that

$$\phi'' \sim A \exp\left(-\frac{1}{2}\xi^2\right),$$

and that its solution is

$$\phi \sim A\xi^{-2} \exp\left(-\frac{1}{2}\xi^2\right) \tag{4.72}$$

The constant β has been obtained as

$$\beta = \lim_{\eta \to \infty} (\eta - f) = 1,21678 \tag{4.73}$$

Thus an approximate formula for $f(\eta)$, valid for large η, is

$$f(\eta) \sim \eta - \beta + A(\eta - \beta)^{-2} \exp\left(-\frac{1}{2}(\eta - \beta)^2\right). \tag{4.74}$$

An immediate consequence of this result is that the transverse velocity component v is given by the asymptotic formula

$$\frac{v}{U_0} \sim \beta(2R_x)^{-\frac{1}{2}}. \tag{4.75}$$

Thus, v does not tend to zero, and this is one of the characteristic features of boundary-layer solutions.

4.8.3 Steady flow in the boundary layer along a cylinder near the forward stagnation point

We consider in this section a two-dimensional flow around a cylindrical obstacle, with a boundary layer extending from the front stagnation point in both directions around the cylinder. Sufficiently near $x = 0$, the velocity U just outside the boundary layer may be represented by the formula $U = U_1 \frac{x}{\ell}$, where U_1 is a velocity and ℓ is a length determined by the overall flow; x is measured along the surface from the stagnation point, see Fig. 4.4. This problem was also dealt with by Blassius

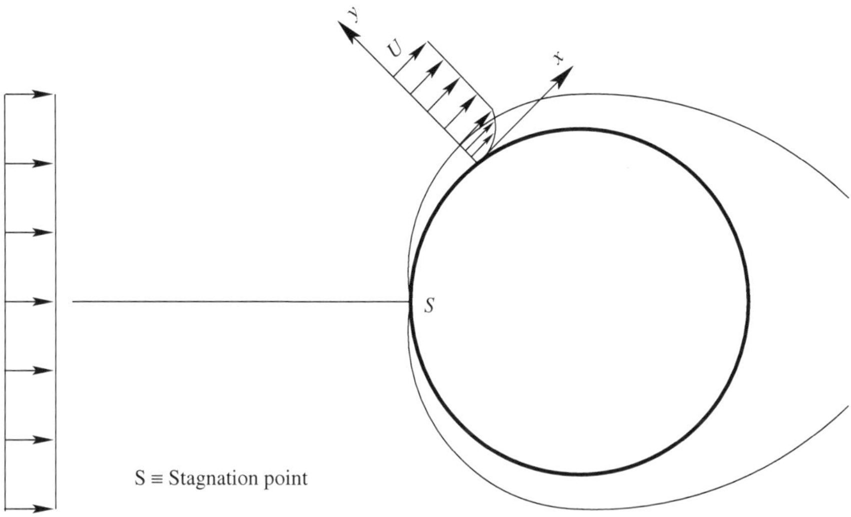

Figure 4.4: The schematic description of boundary layer along a cylinder.

in 1908. The equations of motion and continuity (4.42) and (4.43) in steady state flow, then become

$$u\frac{\partial u}{\partial x} + v\frac{\partial u}{\partial y} = U_1\frac{x}{\ell} + \nu\frac{\partial^2 u}{\partial y^2}, \tag{4.76}$$

$$\frac{\partial u}{\partial x} + \frac{\partial v}{\partial y} = 0, \tag{4.77}$$

The boundary conditions needed here are:

$$y = 0, \quad u = v = 0; \quad x = 0, \quad u = 0; \quad y = \infty, \quad u = U_1\left(\frac{x}{\ell}\right). \tag{4.78}$$

As before, we can look for a similarity solution in which

$$\psi = \left(\frac{\nu U_1}{\ell}\right)^{\frac{1}{2}} x f(\eta),$$

$$\eta = \left(\frac{U_1}{\nu\ell}\right)^{\frac{1}{2}} y = R^{\frac{1}{2}}(y/\ell).$$

The differential equation in the non-dimensional form with the boundary conditions given by

$$f''' + ff'' - f'^2 + 1 = 0, \tag{4.79}$$

$$f(0) = f'(0) = 0, \qquad f'(\infty) = 1 \tag{4.80}$$

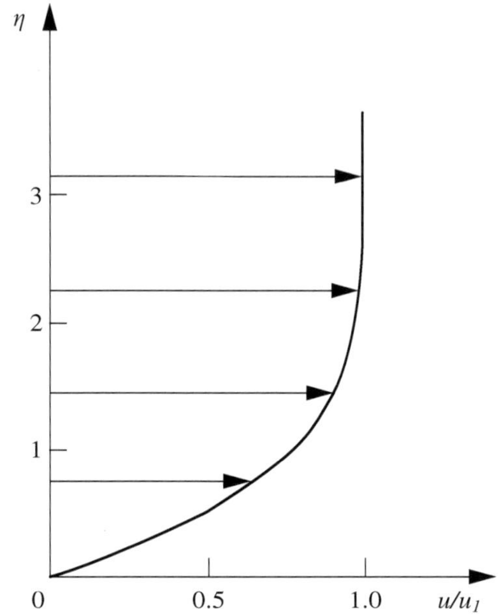

Figure 4.5: The Hiemenz velocity profile (from [1]).

We also have the velocity components as

$$u = (U_1 x/\ell) f'(\eta) = U f'(\eta)$$
$$v = -(\nu U_1/\ell)^{\frac{1}{2}} f(\eta).$$

The equation (4.79) is usually known as the Hiemenz (1911) equation, although it was originally derived by Blasius. The method of solution is that described for the Blasius equation, and the solution for the velocity profile $f'(\eta)$ is depicted here in Fig. 4.3.

Remark

Given that $f''(0) = 1.23259$ for the Hiemenz solution, and that $(\eta - f) \rightarrow 0.64790$, the following quantities can be deduced:

$$\tau_w = 1.23259 \mu U_1 (U_1/\nu\ell)^{1/2} x/\ell,$$
$$\delta_1 = 0.64790(\nu\ell/U_1)^{1/2},$$
$$\delta_2 = 0.29234(\nu\ell/U_1)^{1/2}.$$

We note that the momentum and displacement thicknesses of the boundary layer are not zero $x = 0$, but that they are constant. The reason for this is that the tendency to thicken with increasing x due to viscous diffusion is precisely balanced by the tendency to thin in the presence of a favourable pressure gradient.

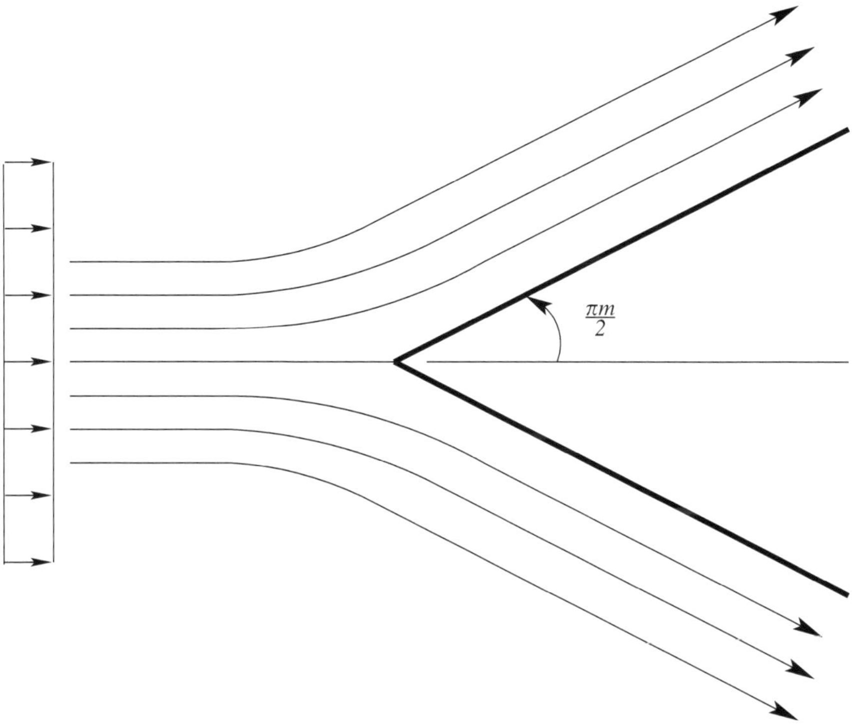

Figure 4.6: The flow around a wedge

4.8.4 Steady flow along a wedge: the Falkner-Skan solutions

Falkner and Skan have noted that the solutions of the preceding problems are special cases of a class of solutions for which

$$U(x) = U_m(x/\ell)^m,$$

this external velocity being attained for the values of $0 \leq m \leq 1$ in the vicinity of the apex of a wedge of semi-angle $\pi m/2$ upon which a uniform stream impinges symmetrically, see Fig. 4.6.

We introduce a stream function

$$\psi = \left(\frac{2U\nu x}{m+1}\right)^{1/2} f(\eta),$$

$$\eta = \left\{\frac{(m+1)U}{2\nu x}\right\}^{1/2} y,$$

and after deriving the appropriate values of u, v, etc., and submitting these into the momentum equation, it is found that

$$f''' + ff'' + \frac{2m}{m+1}\left(1 - (f')^2\right) = 0, \tag{4.81}$$

with boundary conditions

$$f(0) = f'(0) = 0, \quad f'(\infty) = 1.$$

This particular form of the equation was first given by Hartree, the equation obtained by Falkner and Skan differing therefrom to the extent of a linear transformation. It is found that (4.81) has a unique solution when m is positive, but for negative values of m there is more than one solution. Hartree calculated a range of solutions for different values of m, including $m = 1$ (the Hiemenz solution), $m = 0$ (the Blasius solution) and for negative m decreasing to $m = -0.0904$. For this last value of m, it is found that

$$f''(0) = 0,$$

so that $\tau_w = 0$ for all x. This solution therefore represents a flow whose external velocity forces the boundary layer to be everywhere on the point of separation.

Remark

It is worth mentioning that the Falkner-Skan solutions are very valuable as special cases, for each of which the solutions of the boundary layer equations depend on the solution of a single ordinary differential equation, which can readily be obtained with great accuracy. On the other hand, the solutions are of limited physical significance. The Blasius solution, strictly speaking, is appropriate to flow in an unbounded medium into which is introduced a plate of zero thickness. The Hiemenz solution, although an exact solution of the Navier-Stokes equation, is strictly related to flow towards an infinite plane, and for a finite body is correct only near to the forward stagnation point. The Falkner-Skan solutions are valid near the apex of the wedge of semi-angle $\pi m/2$, which immediately creates difficulties when m is negative. We do not want to pursue this matter any further. Interested readers are referred to Curle and Davies [1]. In the following section, we demonstrate the Pohlhausen's method of solving boundary layer equations.

4.9 Pohlhausen's method

Pohlhausen's method seems to be a practical method in solving boundary layer equations. The methods discussed in the previous sections, although extremely accurate, but it is most unlikely that they will, in themselves, be of any practical use. The idea used by Pohlhausen consists essentially of making some sort of plausible assumptions concerning the general shape of the velocity profile. Thus we write

$$\frac{u}{U} = f(\eta, \epsilon_1, \epsilon_2, \epsilon_3, \cdots) \tag{4.82}$$

where $\eta = y/\delta$, and $\epsilon_1, \epsilon_2, \epsilon_3$, etc. are parameters to be determined; δ is the characteristic length of boundary layer thickness. The greater accuracy depends on the number of parameters we use; the more parameter we use the more accuracy we can expect. For simplicity, Pohlhausen used a one parameter method, in which chose

$$\frac{u}{U} = f(\eta) = a_0 + a_1\eta + a_2\eta^2 + a_3\eta^3 + a_4\eta^4. \tag{4.83}$$

The boundary conditions which are satisfied by the true velocity profile are as follows. At the edge of the boundary layer, $y \to \infty$, we have $u \to U$, and hence

$$\frac{\partial u}{\partial y}, \frac{\partial^2 u}{\partial y^2}, \frac{\partial^3 u}{\partial y^3}, \cdots \to 0.$$

At the wall $y = 0$, the primary boundary conditions are

$$u = 0, \quad v = 0.$$

Secondary boundary conditions are given by the momentum equation, namely,

$$\mu\left(\frac{\partial^2 u}{\partial y^2}\right)_w = \frac{dp}{dx} = -\rho U\frac{dU}{dx},$$

$$\text{and} \quad \left(\frac{\partial^3 u}{\partial y^3}\right)_w = 0.$$

The coefficients in (4.83) are chosen by Pohlhausen to satisfy the conditions

$$u = 0, \quad \frac{\partial^2 u}{\partial y^2} = -\frac{U}{\mu}\frac{dU}{dx}, \quad \text{when} \quad y = 0$$

$$\text{and} \quad u = U, \quad \frac{\partial u}{\partial y} = 0, \quad \frac{\partial^2 u}{\partial y^2} = 0, \quad \text{when} \quad y = \delta.$$

Example 4.1

By direct substitution, it can be shown that when Pohlhausen's boundary conditions are used to determine the coefficients, (4.83) becomes

$$\frac{u}{U} = f(\eta) = 2\eta - 2\eta^3 + \eta^4 + \frac{1}{6}A\eta(1-\eta)^3,$$

where $A = \frac{\delta^2}{\nu}\frac{dU}{dx}$.

Details are given below.

Solution

The dimensionless form of the solution of the boundary layer equation is assumed in quartic polynomial as

$$\frac{u}{U} = f(\eta) = a_0 + a_1\eta + a_2\eta^2 + a_3\eta^3 + a_4\eta^4.$$

where $\eta = \frac{y}{\delta}$. This solution has five unknown constants, namely, a_0, a_1, a_2, a_3, and a_4. Therefore we need five boundary conditions to determine these five constants explicitly. According to Pohlhausen we can have these five boundary conditions, two at the plate wall $y = 0$ and the other three at the edge of the boundary layer $y = \delta(x)$. They can be written systematically in the non-dimensional form as follows.

$$\eta = 0: \qquad f(0) = 0; \qquad f''(0) = -\left(\frac{\delta^2}{\nu}\frac{dU}{dx}\right) = -A$$

$$\eta = 1: \qquad f(1) = 1; \qquad f'(1) = f''(1) = 0.$$

Using these five conditions we get five algebraic equations to determine these five constants. It is easy to verify that these constants are found as

$$a_0 = 0,$$

$$a_1 = 2 + \frac{A}{6},$$

$$a_2 = -\frac{A}{2},$$

$$a_3 = \frac{A}{2} - 2,$$

$$\text{and } a_4 = 1 - \frac{A}{6}.$$

Thus using these values in the solution above, and after a little reduction we obtain exactly

$$\frac{u}{U} = f(\eta) = 2\eta - 2\eta^3 + \eta^4 + \frac{1}{6}A\eta(1-\eta)^3.$$

If we use the cubic polynomial such as

$$\frac{u}{U} = f(\eta) = a_0 + a_1\eta + a_2\eta^2 + a_3\eta^3$$

with four boundary conditions; two at the wall namely $f(0) = 0$ and $f''(0) - A$ and the other two at the edge of the boundary layer, namely $f(1) = 1$ and $f'(1) = 0$, we obtain

$$a_0 = 0, \; a_1 = \frac{3}{2} + \frac{A}{4}, \; a_2 = -\frac{A}{2}, \; a_3 = -\frac{1}{2} + \frac{A}{4}.$$

The solution can be written as

$$\frac{u}{U} = f(\eta) = \frac{1}{2}(3\eta - \eta^3) + \frac{A}{4}\eta(1 - \eta)^2.$$

It will be hard to see how accurate these two solutions are unless tested with the experimental data. However, we have to appreciate the efforts made by Pohlhausen to determine the solution in a very simple way for such a complicated problem.

Now it can be easily deduced using the qartic polynomial solution that

$$\begin{aligned}
\tau &= \frac{\mu U}{\delta}\left(1 + \frac{1}{6}A\right),\\
\delta_1 &= \delta\left(\frac{3}{10} - \frac{A}{120}\right),\\
\delta_2 &= \frac{1}{315}\delta\left(37 - \frac{A}{3} - \frac{5}{14}A^2\right).
\end{aligned} \tag{4.84}$$

By substituting into the momentum integral (4.52), it can be easily shown that the resulting equation for δ (or A) may be written in the form

$$\frac{dZ}{dx} = \frac{g(A)}{U} + Z^2U''h(A),$$

$$\text{or} \quad \frac{dA}{dx} = g(A)\frac{U'}{U} + (A + A^2h(A))\frac{U''}{U'},$$

where $Z = \frac{\delta^2}{\nu}$ and $g(A)$, $h(A)$, are functions of A.

Concluding remarks

It is recognized that Pohlhausen's method appears to give good results in region of favourable pressure gradient. However, there are two criticisms that may be levelled in his method. The first concerns the arbitrary choice as to which boundary conditions shall be satisfied by the approximate velocity profile chosen. For a simple instance, Pohlhausen, using a quartic polynomial, satisfies three conditions at the edge of the boundary-layer, and two conditions at the wall. Suppose, instead, he had chosen either a cubic velocity-profile and rejected one of the boundary conditions at the wall, or a quintic profile, satisfying a further condition at the wall. It is hard to predict the accuracy of the method unless the results are compared with available experimental data. It is advisable therefore, to see the accuracy of the method, use a cubic or quintic profile, satisfying the appropriate boundary conditions.

A second, and very practical, criticism of Pohlhausen method is that U'', the second derivative of the external velocity U, appears explicitly in the formulation. If $U(x)$ is given analytically, this hardly matters, but if $U(x)$ has been obtained by experiment the resulting scatter will make it extremely difficult to obtain more than a very rough estimate of the second derivative. Fortunately, we can avoid this difficulty by introducing the parameter

$$\lambda = U'\frac{\delta_2^2}{\nu} = \left(\frac{\delta_2}{\delta}\right)^2 A.$$

The momentum integral equation may be written as

$$\frac{\tau_w}{\rho} = U^2\frac{d\delta_2}{dx} + UU'(\delta_1 + 2\delta_2)$$

$$\text{or} \quad \frac{\tau_w\delta_2}{\mu U} = \frac{U\delta_2}{\nu}\frac{d\delta_2}{dx} + \frac{U'\delta_2^2}{\nu}\left(\frac{\delta_1}{\delta_2} + 2\right).$$

This can be further reduced to the following.

$$\ell = \frac{U\delta_2}{2\nu}\frac{d\delta_2^2}{dx} + \lambda(H+2)$$

where

$$\ell = \frac{\tau_w\delta_2}{\mu U} \quad \text{and} \quad H = \frac{\delta_1}{\delta_2}. \tag{4.85}$$

Thus we have

$$U\frac{d}{dx}\left(\frac{\delta_2^2}{\nu}\right) = U\frac{d}{dx}\left(\frac{\lambda}{U'}\right) = 2\left(\ell - \lambda(H+2)\right) = L. \tag{4.86}$$

It is fairly easy to use the results (4.84), together with the definition of (4.85) and (4.86), to deduce ℓ, H and L as a function of λ.

Example 4.2

Approximate solutions to the boundary-layer equations may be obtained from the momentum integral equation

$$\frac{\tau_w}{\rho U^2} = \frac{d\delta_2}{dx} + \frac{\delta_1 + 2\delta_2}{U}\frac{dU}{dx},$$

using an approximate velocity-profile

$$\frac{u}{U} = f(\eta) = a + b\sin(\pi\eta/2) + c\sin^2(\pi\eta/2),$$

where $\eta = y/\delta$.

Indicate how suitable values of a, b, c, may be obtained, and thence determine the value of $\lambda = U'\delta_2/\nu$ corresponding to the separation profile of this family.

Solution

There are three arbitrary constants to be determined in this case. Therefore, we need three boundary conditions of the boundary layer equations. These are two wall conditions and one condition at the edge of the boundary layer will be sufficient to determine these three constants.

We know that the boundary conditions are

$$\eta = 0, \qquad f(0) = 0; \qquad f''(0) = -A,$$
$$\eta = 1, \qquad f(1) = 1.$$

The solution is assumed as above and so we calculate the f' and f'':

$$f(\eta) = a + b\sin(\pi\eta/2) + \frac{c}{2}\left[1 - \cos(\pi\eta)\right],$$
$$f'(\eta) = \frac{b\pi}{2}\cos(\pi\eta/2) + \frac{c\pi}{2}\sin(\pi\eta),$$
$$f''(\eta) = -b\left(\frac{\pi}{2}\right)^2\sin(\pi\eta/2) + c\frac{\pi^2}{2}\cos(\pi\eta).$$

Using the given boundary conditions, we obtain the values of $a = 0$, $b = 1 + \frac{2A}{\pi^2}$, and $c = -\frac{2A}{\pi^2}$.

Therefore

$$f(\eta) = \left(1 + \frac{2A}{\pi^2}\right)\sin(\pi\eta/2) - \frac{2A}{\pi^2}\sin^2(\pi\eta/2).$$

To determine the value of $\lambda = U'\delta_2^2/\nu$, we need to evaluate the momentum thickness $\delta_2(x)$ first. So let us calculate this important factor. By definition, we have

$$\begin{aligned}
\delta_2(x) &= \int_0^{\delta} \frac{u}{U}\left(1 - \frac{u}{U}\right)dy \\
&= \delta(x)\int_0^1 f(1-f)d\eta \\
&= \delta(x)\int_0^1 (f - f^2)d\eta \\
&= \delta(x)\left\{\int_0^1 \left[(b\sin(\pi\eta/2) + c\sin^2(\pi\eta/2)) \right.\right. \\
&\qquad \left.\left. - (b\sin(\pi\eta/2) + c\sin^2(\pi\eta/2))^2\right]d\eta\right\}
\end{aligned}$$

Now using the substitution $\pi\eta/2 = \theta$ such that $d\eta = \frac{2}{\pi}d\theta$ with the limits $0 \leq \theta \leq \pi/2$, we evaluate the integrals at once.

Let

$$I_1 = \frac{2}{\pi}\int_0^{\pi/2} (b\sin(\theta) + c\sin^2(\theta))d\theta$$
$$= \frac{2}{\pi}\left(b + c\frac{\pi}{4}\right)$$

Let

$$I_2 = \frac{2}{\pi}\int_0^{\pi/2} (b\sin(\theta) + c\sin^2(\theta))^2 d\theta$$
$$= \frac{2}{\pi}\int_0^{\pi/2} [b^2 sin^2\theta + c^2\sin^4\theta + 2bc\sin^3\theta]d\theta$$
$$= \frac{2}{\pi}\left[\frac{\pi}{4}b^2 + \frac{3\pi}{16}c^2 + \frac{4}{3}bc\right]$$

Hence the momentum thickness $\delta_2(x)$ is given by

$$\delta_2(x) = \frac{2}{\pi}\delta(x)\left\{\left(b + c\frac{\pi}{4}\right) - \left[\frac{\pi}{4}b^2 + \frac{3\pi}{16}c^2 + \frac{4}{3}bc\right]\right\}$$

Once the value of $\delta_2(x)$ has been determined the value of λ in question is obvious.

Example 4.3

If $U = U_0$ (constant), and assuming an approximate velocity-profile

$$u = U\tanh\eta, \quad \eta = y/\delta$$

determine the skin-friction using the momentum integral equation

$$\frac{\tau_w}{\rho U^2} = \frac{d\delta_2}{dx} + \frac{\delta_1 + 2\delta_2}{U}\frac{dU}{dx}.$$

Solution

When the external velocity U is uniform, the momentum integral equation reduces to a simple form

$$\frac{\tau_w}{\rho U^2} = \frac{d\delta_2}{dx}.$$

Thus we need to evaluate only the $\delta_2(x)$. By definition we have

$$\begin{aligned}\delta_2(x) &= \int_0^\infty \frac{u}{U}\left(1 - \frac{u}{U}\right) dy \\ &= \delta(x)\int_0^\infty f(1-f)d\eta \\ &= \delta(x)\int_0^\infty (\tanh\eta - \tanh^2\eta)d\eta\end{aligned}$$

We know $\tanh^2\eta = 1 - sech^2\eta$ and so inserting this relation into the integral, we obtain

$$\begin{aligned}\delta_2(x) &= \delta(x)\int_0^\infty (\tanh\eta - 1 + sech^2\eta)d\eta \\ &= \delta(x)[\ln(\cosh\eta) - \eta + \tanh\eta]_0^\infty \\ &= \delta(x)[\lim_{\eta\to\infty}(\ln(\cosh\eta) - \eta) + 1] \\ &= \delta(x)[1 - \ln 2]\end{aligned}$$

Hence

$$\frac{\tau_w}{\rho U_0^2} = \frac{d\delta_2}{dx} = \left(\frac{1-\ln 2}{2}\right)\left(\frac{U_0 x}{\nu}\right)^{-\frac{1}{2}} = 0.153426\left(\frac{U_0 x}{\nu}\right)^{-\frac{1}{2}}.$$

This is the required skin-friction.

4.10 Flow in laminar wakes and jets

There are many interesting problems that may be studied by means of boundary-layer approximation. It is impossible to investigate many of these problems by the method already discussed in this chapter. We shall consider the following interesting and important physical problem in this section.

Let us consider the wake behind a flat plate, set parallel to a uniform stream. We assume that Blasius boundary layers develop on either side of the plate, as illustrated in Fig. 4.4. Downstream of the plate, the profile is distorted, since viscosity requires it to be rounded at the plane of symmetry. The region of significant distortion diffuses outwards at positions downstream, until a region of fully developed wake flow is reached. We note that, mathematically, the change in profile downstream of the trailing edge is characterized by a change in the boundary conditions from $u = 0$ when $y = 0$, to $\frac{\partial u}{\partial y} = 0$ when $y = 0$.

It is fairly clear that there are two transverse length scales in the adjustment region, one characteristic of the overall wake thickness, and the other of the thickness of the region. Clearly there is little hope of obtaining a simple mathematical solution in this region. In the fully developed wake flow, however, it is fairly straightforward

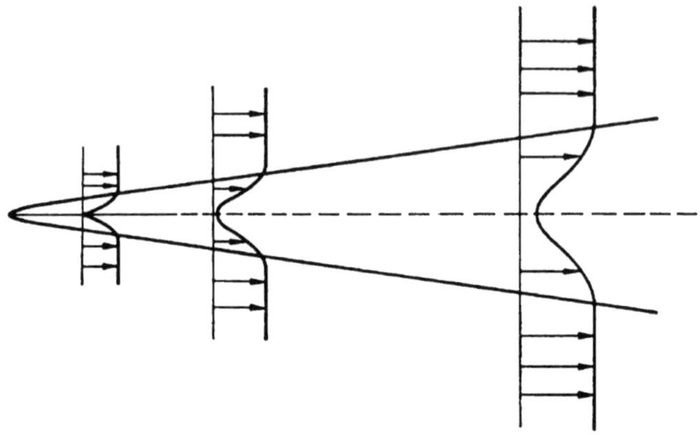

Figure 4.7: Flow in a laminar wake.

to obtain a solution which is valid asymptotically for larger values of x when the wake has become wide and the velocity defect correspondingly small.

Let us assume that the wake spreads slowly, so that the boundary layer approximation holds, and we write

$$u = u_0 + u'$$
$$v = v'$$

where squares, products and higher powers of u', v', are neglected. Then the continuity and momentum equations become

$$\frac{\partial u'}{\partial x} + \frac{\partial v'}{\partial y} = 0,$$

and $$u_0 \frac{\partial u'}{\partial x} = \nu \frac{\partial^2 u'}{\partial y^2}.$$

We now look for a solution in the following similarity form

$$u' = Ax^m f(\eta), \quad \eta = y\left(\frac{u_0}{\nu x}\right)^{1/2}, \tag{4.87}$$

and it is found, after substitution, that

$$f'' + \frac{1}{2}\eta f' - mf = 0, \tag{4.88}$$

with boundary conditions

$$f'(0) = f'''(0) = f^{v}(0) = \cdots = 0,$$

and $$f(\eta) \to 0 \text{ as } \eta \to \infty.$$

The value of m can be found either mathematically or physically. Mathematically we write (4.88) as

$$f'' + \frac{1}{2}\eta f' + \frac{1}{2}f = \left(m + \frac{1}{2}\right) f,$$

so that, integrating with respect to η from 0 to ∞,

$$\left[f' + \frac{1}{2}\eta f\right]_0^\infty = 0 = \left(m + \frac{1}{2}\right) \int_0^\infty f \, d\eta.$$

Hence we have $m + \frac{1}{2} = 0$, i.e., $m = -\frac{1}{2}$, since $\int_0^\infty f \, d\eta$ cannot be zero, being proportional to the flux of mass defect in the wake.

Alternatively, the flux of momentum defect in the wake must be constant, since there is no pressure gradient, as may be shown by integrating the momentum equation from $y = 0$ to $y = \infty$. Thus, after linearizing, we find that

$$\int_0^\infty u' dy = \text{constant}.$$

But, (4.87),

$$\int_0^\infty u' dy \propto x^{m+\frac{1}{2}} \int_0^\infty f \, d\eta,$$

and this can only be constant when $m = -\frac{1}{2}$, the value given above.

With this value of m, (4.88) becomes

$$f'' + \frac{1}{2}\eta f' + \frac{1}{2}f = 0,$$

which integrates twice to yield

$$f((\eta) \propto \exp\left(-\frac{1}{4}\eta^2\right).$$

Hence (4.87) becomes

$$u' = Ax^{-\frac{1}{2}} \exp\left(-\frac{1}{4}\eta^2\right), \quad \eta = y\left(\frac{u_0}{\nu x}\right)^{1/2}, \tag{4.89}$$

for some constant A. To calculate A we note that the momentum thickness of half of the wake is given, after linearizing, by

$$\begin{aligned} \delta_2 &= -\int_0^\infty \frac{u'}{u_0} dy \\ &= -A\left(\frac{\pi \nu}{u_0^3}\right)^{1/2}. \end{aligned}$$

Being constant along the wake, this equals its value at the trailing edge, so by setting $x = \ell$ in $\delta_2 = 0.66412(\frac{\nu x}{u_0})^{1/2}$, we find that

$$
\begin{aligned}
A &= -\left(\frac{u_0^2}{\pi \nu}\right)^{1/2} \delta_2 \\
&= -\frac{0.66412}{\pi^{1/2}} u_0 \ell^{1/2}.
\end{aligned}
$$

Finally, (4.89) then becomes

$$
u' = u_0 \left\{ 1 - \frac{0.66412}{\pi^{1/2}} \left(\frac{\ell}{x}\right)^{1/2} \exp\left(-\frac{u_0 y^2}{4\nu x}\right) \right\}.
$$

Example 4.4

Repeat the above analysis for the analogous problem of flow in a two-dimensional laminar jet, far from the orifice.

Solution

We assume that there is no pressure gradient along the jet, and use the boundary layer approximation. Then the equations become

$$
u\frac{\partial u}{\partial x} + v\frac{\partial u}{\partial y} = \nu \frac{\partial^2 u}{\partial y^2}
$$

$$
\frac{\partial u}{\partial x} + \frac{\partial v}{\partial y} = 0.
$$

We look for a similar solution in the following form

$$
\psi = u_0 x \left(\frac{u_0 x}{\nu}\right)^m f(\eta),
$$

where $$\eta = \frac{y}{x}\left(\frac{u_0 x}{\nu}\right)^n.$$

Substitution and simplification leads to

$$
\left(\frac{u_0 x}{\nu}\right)^{-m+n-1} f''' + (m+1) f f'' - (m+n) f'^2 = 0.
$$

For consistency $n = m + 1$, and so

$$
f''' + (m+1) f f'' - (2m+1) f'^2 = 0.
$$

The value of m then follows either mathematically or physically as before, and this yields $m = -\frac{2}{3}$, so that the differential equation becomes

$$
f''' + \frac{1}{3} f f'' + \frac{1}{3} f'^2 = 0.
$$

Hence after two integrations, we find that

$$f = a \tanh\left(\frac{1}{6}a\eta\right),$$

where the constant a may be related to the momentum flux in the jet. It can be shown that the maximum velocity in the jet varies as $x^{-1/3}$, and the mass flux in the jet varies as $x^{1/3}$, so that the jet must entrain air.

For further information on this topic, the reader is referred to the work of Jen, Y. [2] to Rahman, M. [6] and Sarpkaya, T. [8] to Stoker, J.J. [11] as listed in the reference section.

4.11 Exercises

1. Assume that the local velocity $u(x, y)$ differs slightly from the value $U(x)$ at the edge of the boundary layer, then linearize the boundary layer equation in the manner of Oseen to obtain

$$U\frac{\partial u}{\partial x} + u\frac{dU}{dx} = 2U\frac{dU}{dx} + \nu\frac{\partial^2 u}{\partial y^2}.$$

   Seek a similar solution of this equation, for which the external velocity U is $U = \beta_m \xi^m$, $\xi = x/c$, for some value of m. By choosing $u = Uf'(\eta)$ for suitably chosen η, show that f satisfies the equation of the form

$$f''' + 2m(1 - f') - \frac{1}{2}(m-1)\eta f'' = 0.$$

   By obtaining, in effect, the equivalent momentum integral equation, deduce that the resulting boundary layer is everywhere on the point of separation if $m = -\frac{1}{3}$.
2. Assume that the boundary layer approximation may be applied, obtain the equation governing flow in the wake behind a flat plate at zero incidence to a uniform stream velocity U_0. Deduce, from first principles that the momentum integral equation takes the form

$$\delta_2 = \int_0^\infty \frac{u}{U_0}\left(1 - \frac{u}{U_0}\right)dy = 0.664\left(\frac{\nu\ell}{U_0}\right)^{1/2},$$

   where ℓ is the length of the plate and nu is the kinematic viscosity of the fluid.
3. Prove that $W = A cos\ k(z + ih - ct)$ is the complex potential for the propagation of simple harmonic surface waves of small height on water of depth h, the origin being in the undisturbed free surface. Express A in terms of the amplitude a of the small oscillations. Prove that $c^2 = \frac{g}{k}\ tanh kh$, and deduce that every value of V less than $(gh)^{\frac{1}{2}}$ is the velocity of some wave. Prove that each particle describes an elliptical path about its equilibrium position. Obtain the corresponding result when the water is infinitely deep (Note: $z = x + iy$).

4. Calculate the kinetic and potential energies associated with a single train of progressive waves on deep water, and, from the condition that these energies are equal, obtain the formula $c^2 = (gL/2\pi)$. Show how this result is modified when the wavelength is so small that the potential energy due to surface tension is not negligible.

References

[1] Curle, N. and Davies, H.J., *Modern Fluid Dynamics, Vol. 1: Incompressible Flow*, D. Van Nostrand Company Ltd.: London, 1968.

[2] Jen, Y., Wave refraction near San Pedro Bay, California. *J. Waterways Harbors and Coastal Engg. Div.*, **95(WW3)**, pp. 379–393, 1969.

[3] Keulegan, G.H. and Harrison, J., Tsunami refraction diagrams by digital computer. *J. Waterways Harbors and Coastal Engg. Div.*, **96(WW2)**, pp. 219–233, 1970.

[4] Mei, C.C., *The Applied Dynamics of Ocean Surface Waves*, Wiley: New York, 1983.

[5] Rahman, M., *The Hydrodynamics of Waves and Tides, with Applications*, Computational Mechanics Publications: Southampton, 1988.

[6] Rahman, M., *Applied Differential Equations for Scientists and Engineers, Vol. 1: Ordinary Differential Equations*, Computational Mechanics Publication: Southampton, 1991.

[7] Rosenhead, L., ed., *Laminar Boundary Layers*, Oxford University Press: Oxford, 1963.

[8] Sarpkaya, T. and Isaacson, M., *Mechanics of Wave Forces on Offshore Structures*, Van Nostrand Reinhold Company: New York, 1981.

[9] Skovgaard, O., Jonsson, I. and Bertelsen, J. Computation of wave heights due to refraction and friction. *J. Waterways Harbors and Coastal Engg. Div.*, **101(WW1)**, pp. 15–32, 1975.

[10] Sommerfeld, A., *Partial Differential Equations in Physics*, Academic Press: New York, 1949.

[11] Stoker, J.J., *Water Waves*, Interscience: New York, 1957.

[12] Rahman, M., *Advanced Vector Analysis for Scientists and Engineers*, WIT Press: Southampton, UK, 2007.

CHAPTER 5

Similarity analysis in fluid flow

Professor Carlos A Brebbia

Professor Carlos Brebbia graduated at the University of Litoral, Argentina. He received his PhD in Civil Engineering at Southampton University, England in 1968 and worked at MIT, USA and the Central Electrical Research Laboratory, UK as well as the University of Princeton, USA.

Professor Carlos Brebbia was appointed as Lecturer at the University of Southampton, UK, in 1969 and as senior lecturer in 1976 before becoming a Reader in Computational Engineering. In 1979, he became Professor of Civil Engineering at the University of California at Irvine. He is now Chairman of Computational Mechanics and Director of Wessex Institute of Technology.

Professor Carlos Brebbia has written over 250 scientific papers, is author or co-author of 13 books and editor or co-editor of over 80 other books. He is an internationally well-known specialist in numerical methods, finite and boundary elements and the computer solution of engineering problems. Professor Carlos Brebbia is the Editor or co-editor of five scientific journals and has carried out a great deal of consultancy work for the engineering industry as well as having participated in many international projects and conferences.

This chapter is dedicated to Professor Carlos Brebbia, the Founding Director of the Wessex Institute of Technology, Ashurst Lodge, Ashurst, Southampton, UK. He introduced the Boundary Element Method (known as BEM) to solve engineering problems. For a couple of years he was in the University of California, USA as Professor and then he left for England to create Wessex Institute of Technology (WIT) in Ashurst for the post-graduate students. He created the WIT Press for publication of high-level research books and international conference papers. He dedicated his entire life for higher education and learning to the benefit of young engineers and scientists.

Because of his devotion to the higher learning and scholarship, I was very impressed at his scientific endeavour and was associated with the institute since 1980, and now I am a member of the Governing Council of WIT. Professor Brebbia has published numerous scientific papers and books as mentioned above. It is worth noting that he organizes every year about 30 international conferences in various engineering subjects in different cities in the world. Through his service, many graduate students, scientists and engineers around the world derive benefits. It is my pleasure to dedicate this chapter of my fluid mechanics memoir *to this elite person for his unparallel, unmatched, unique and extraordinary innovative contributions to the world of science and engineering and also for his loving care to the higher learning and scholarship.*

5.1 Introduction

In this chapter we discuss a very important fluid flow problem of physical interest. In the last chapter we devoted significantly to the development of boundary layer theory in two-dimensional fluid flow. As we have seen in the previous chapter, considerable amount of knowledge about similar solutions is needed and so we considered the fundamentals of the similarity technique and how this method can be used in real physical problems. We discussed the problem of natural convection flow with diffusion and chemical reactions in an infinite vertical plate and the solutions are obtained using the similarity technique. The main idea of this method is to reduce the dimension of the problem by one and then use analytical procedure or numerical method to get the solution in the similarity variables. This is exactly what we have done in this problem.

5.2 Concept and definition of heat and mass diffusion

Boundary value problems involving the principles of heat and mass diffusion in a fluid medium, where the results are directly influenced by the process of fluid motion may in general be termed convection process. If the motion of the fluid is determined by the boundary conditions specified externally to the system, such as forcing air through a passage in which an external pressure gradient is specified, the process is called forced convection. Otherwise, if the fluid velocities are caused by the effects of gravity force, i.e. by the interaction of a body force with variable density arising from heat or mass diffusion, then the process is called natural convection.

The phenomena of natural convection can be observed in the atmosphere, in bodies of water, adjacent to domestic heating radiators or over sun-heated fields and roads. The basic equations which describe natural convection flows are similar to those of other fluid flows and diffusion processes, with the essential difference that, in natural convection, the motion of a fluid arises from buoyancy and not from an imposed motion or pressure gradient. The fundamental equations and their boundary layer approximations for natural convection flows are derived in the following section.

The characteristic numbers of natural convection flows which are commonly used are as follows:

Local Grashof number: $$Gr_x = \frac{g\beta * (c_0 - c_\infty)x^3}{\nu^2}$$

Local Nusselt number: $$Nu_x = \frac{J''(x)}{(c_0 - c_\infty)x^3}\left(\frac{x}{D}\right)$$

Local reaction rate number: $$\epsilon(x) = \frac{2k(c_0 - c_\infty)^{n-3/2}}{\sqrt{g\beta^*}}x^{1/2}$$

Schmidt number: $$Sc = \frac{\nu}{D}$$

Prandtl number: $$Pr = \frac{\nu}{\alpha}$$

where g is the acceleration due to gravity, β^* is the volumetric coefficient of expansion with concentration species, c_0 is the species concentration at the plate surface, c_∞ is the species concentration at the ambient fluid, x is the coordinate length along the plate, ν is the kinematic viscosity of the fluid, D is the chemical molecular diffusivity, α is the thermal molecular diffusivity, $J''(x)$ is the local mass flux, k is the chemical reaction-rate constant and n is the order of reaction.

The quantity

$$Gr = \frac{g\beta^*(c_0 - c_\infty)L^3}{\nu^2}$$

where L is a characteristic length of the plate, is known as the Grashof number. This number is a measure of the vigor of the flow induced due to buoyancy-effects. It arises in the forced momentum balance as the ratio of the relative magnitudes of viscous force and the inertia force. The Grashof number of natural convection flow is analogous to the Reynolds number of forced flow, $\sqrt{Gr}$ corresponds to R. The characteristic feature of this number is that stability and transition are defined in the limits of this number. The quantities $Pr = \frac{\nu}{\alpha}$ and $Sc = \frac{\nu}{D}$ are defined as the Prandtl and Schmidt numbers, respectively. The former one occurs in energy equation whereas the latter one occurs in mass-diffusion equation. These numbers are the indicators of the steepness of the gradient of temperature and concentration, respectively, in the flow field. As for instance, for high Prandtl number fluids, the temperature gradient is much steeper than that for low Prandtl number fluids.

The local mass flow per unit area (mass flux) from the surface of the plate to the fluid may be calculated from Fick's first law of diffusion:

$$\mathbf{J} = -D\nabla c,$$

which is the three-dimensional form of Fick's law. For a two-dimensional case, if we assume that the mass flux component parallel to the surface is very small compared to that perpendicular to the surface, then the mass flux from the plate to the fluid may be written as

$$J''(x) = -D\left(\frac{\partial c}{\partial y}\right)_{y=0}.$$

It is customary to express mass transfer characteristics in terms of a *mass transfer coefficient* h, defined as the mass transfer per unit area (i.e., mass flux) divided by the concentration difference causing the mass-transfer. Since the flux is often variable over the surface even for a uniform concentration difference, the mass transfer coefficient h varies over the surface. Therefore, one may speak of local values h_x or of average values h. This local value is given by

$$h_x = \frac{J''(x)}{c_0 - c_\infty} = -\frac{D}{c_0 - c_\infty}\left(\frac{\partial c}{\partial y}\right)_{y=0}.$$

Multiplying through by $\frac{x}{D}$, a dimensionless combination is found which is called the local Nusselt number (a mass transfer parameter):

$$Nu_x = \frac{h_x x}{D}.$$

The average mass transfer coefficient h from $x = 0$ to ℓ may be determined from the following integral:

$$h = \frac{\text{average mass flux}}{\text{average concentration difference}} = \frac{\left(\frac{1}{\ell}\int_0^\ell J''(x)dx\right)}{\left(\frac{1}{\ell}\int_0^\ell (c_0 - c_\infty)dx\right)}.$$

Thus the average transport parameter is given by

$$Nu_\ell = \frac{h\ell}{D},$$

which is a measure of mass flow per unit area from the surface of the plate to the fluid in dimensionless form. The quantity

$$\frac{2k(c_0 - c_\infty)^{n-3/2}x^{1/2}}{\sqrt{g\beta^*}}$$

arises as an indication of the relative importance of the chemical reaction-rate in the mass diffusion equation and is called the reaction-rate number.

5.3 General statement of the problem

We consider a steady two-dimensional laminar viscous flow over a semi-infinite vertical plate immersed in an ambient fluid. It is assumed that a homogeneous isothermal irreversible chemical reaction of nth order takes place between the chemical constituents of plate and fluid.

Further, we assume the following: the fluid is Newtonian; two-dimensional laminar steady flow is considered; the physical properties associated with the problem such as viscosity, diffusivity, etc. are constant; the Boussinesq approximation is taken into consideration for buoyancy effects, which implies the small density changes in the gravitational field; the reaction number is assumed small; static pressure gradients arising from the convection currents are neglected; inertial forces in the convective field are assumed to be in balance with buoyancy and viscosity forces; and there is no externally applied pressure gradients.

A change in chemical composition of the fluid near the surface of the plate is considered to produce lighter fluid there, which rises as the buoyancy force overcomes gravity force resulting in an upward moving particle. If a heavier fluid were produced the reverse effect would be observed but the mathematical problem be essentially the same. The chemical species is first transferred from the plate to the ambient fluid by diffusion and then carried away by induced convection currents. A distinguishing feature of this problem is that when a chemical reaction occurs in the bulk of the fluid, the diffusing species may be depleted, whereas in problems without chemical reaction, no such effect is possible. The basic equations which govern this process derive from the following principles:

(i) conservation of mass (fluid);
(ii) conservation of momentum;
(iii) conservation of mass (species)- i.e., Fick's second-law of diffusion; and
(iv) appropriate law of chemical reaction.

The boundary layer equations governing this chemical diffusion problem are given by

$$\frac{\partial u}{\partial x} + \frac{\partial v}{\partial y} = 0, \tag{5.1}$$

$$u\frac{\partial u}{\partial x} + v\frac{\partial u}{\partial y} = \nu\frac{\partial^2 u}{\partial y^2} + g\beta^* C, \tag{5.2}$$

$$u\frac{\partial C}{\partial x} + v\frac{\partial C}{\partial y} = D\frac{\partial^2 C}{\partial y^2} - kC^n, \tag{5.3}$$

where $C = c - c_\infty$. The relevant boundary conditions are:

$$y = 0, \quad u = v = 0, \quad \text{for all } x \quad \text{(no slip condition)}, \tag{5.4}$$

$$y \to \infty \quad u = v = 0, \quad \text{for all } x \quad (\text{uniformity at } \infty). \tag{5.5}$$

$$y \to \infty \quad C \to 0, \quad \text{for all } x \quad (\text{uniformity at } \infty). \tag{5.6}$$

Further, $C = C_0(x)$ is prescribed at the plate surface $y = 0$. Here $C_0(x)$ is a given function of x representing the concentration distribution prescribed along the plate. The velocity field and the concentration distribution are illustrated in Fig. 5.1 and 5.2.

5.4 Similarity analysis of the basic equations

Introducing the stream function $\psi(x, y)$ defined by $u = \frac{\partial \psi}{\partial y}$, and $v = -\frac{\partial \psi}{\partial x}$ into the above boundary value problem we obtain

$$\frac{\partial \psi}{\partial y}\frac{\partial^2 \psi}{\partial y \partial x} - \frac{\partial \psi}{\partial x}\frac{\partial^2 \psi}{\partial y^2} = \nu\frac{\partial^3 \psi}{\partial y^3} + g\beta^* C, \tag{5.7}$$

$$\frac{\partial \psi}{\partial y}\frac{\partial C}{\partial x} - \frac{\partial \psi}{\partial x}\frac{\partial C}{\partial y} = D\frac{\partial^2 C}{\partial y^2} - kC^n, \tag{5.8}$$

with boundary conditions

$$y = 0: \quad \frac{\partial \psi}{\partial y} = \frac{\partial \psi}{\partial x} = 0,$$

$$y = 0: \quad C = C_0(x), \tag{5.9}$$

$$y \to \infty: \quad \frac{\partial \psi}{\partial y} = \frac{\partial \psi}{\partial x} = 0,$$

$$y \to \infty: \quad C = 0. \tag{5.10}$$

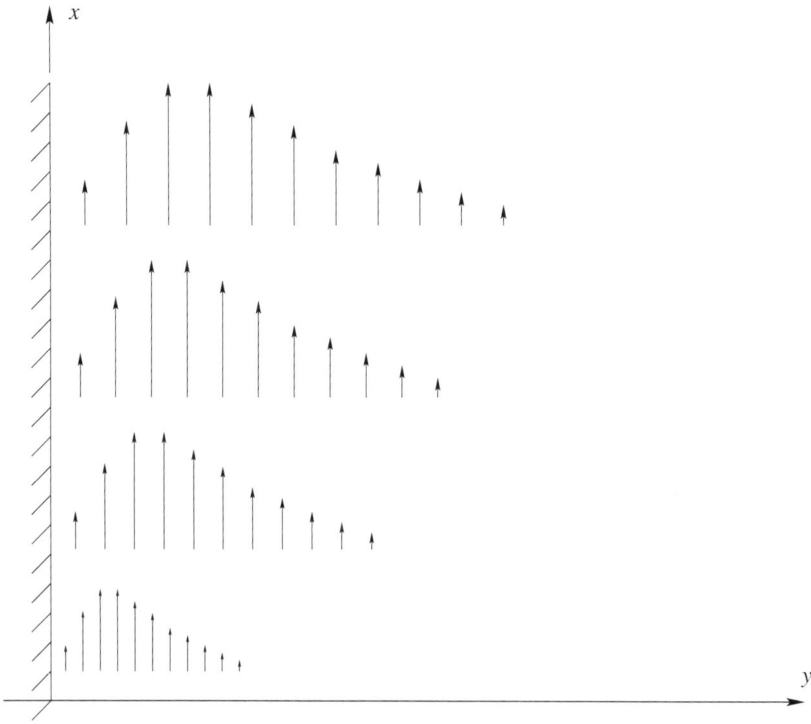

Figure 5.1: Physical plane of velocity field in natural convention.

Consider the following one-parameter transformation group,

$$\begin{aligned}\hat{x} &= a^m x\\ \hat{y} &= a^l y\\ \hat{\psi} &= a^p \psi\\ \hat{C} &= a^q C\end{aligned} \tag{5.11}$$

where a is a parameter and the exponents m, l, p and q are all constants. Then we have the following quantities:

$$\begin{aligned}\psi_y &= a^{-p+l}\ \hat{\psi}_{\hat{y}},\\ \psi_{yy} &= a^{-p+2l}\ \hat{\psi}_{\hat{y}\hat{y}},\\ \psi_{yyy} &= a^{-p+3l}\ \hat{\psi}_{\hat{y}\hat{y}\hat{y}},\\ \psi_x &= a^{-p+m}\ \hat{\psi}_{\hat{x}},\end{aligned}$$

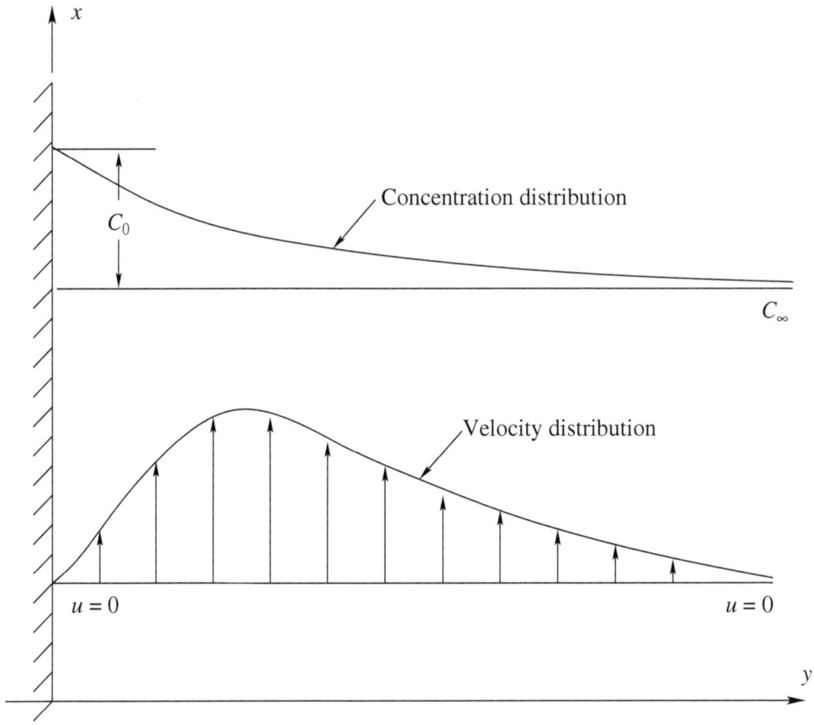

Figure 5.2: Velocity and concentration distributions in the physical plane.

$$
\begin{aligned}
\psi_{yx} &= a^{-p+m+l}\ \hat{\psi}_{\hat{y}\hat{x}},\\
C_x &= a^{-q+m}\ \hat{C}_{\hat{x}},\\
C_y &= a^{-q+l}\ \hat{C}_{\hat{y}},\\
C_{yy} &= a^{-q+2l}\ \hat{C}_{\hat{y}\hat{y}}.
\end{aligned}
\tag{5.12}
$$

Here the subscript notations are meant for the partial derivatives. Substituting (5.12) into the equations (5.7) and (5.8), after a little reduction we obtain

$$\hat{\psi}_{\hat{y}}\hat{\psi}_{\hat{y}\hat{x}} - \hat{\psi}_{\hat{x}}\hat{\psi}_{\hat{y}\hat{y}} = \nu a^{p-m+l}\hat{\psi}_{\hat{y}\hat{y}\hat{y}} + a^{-q+2p-2l-m} g\beta^* \hat{C} \tag{5.13}$$

$$\hat{\psi}_{\hat{y}}\hat{C}_{\hat{x}} - \hat{\psi}_{\hat{x}}\hat{C}_{\hat{y}} = Da^{l-m+p}\hat{C}_{\hat{y}\hat{y}} - ka^{-nq+p+q-l-m}\hat{C}^n \tag{5.14}$$

It (5.13) and (5.14) are invariant under the transformation group (5.11), then we must have

$$
\begin{aligned}
p &= m - l\\
q &= m - 4l\\
\frac{m}{l} &= \frac{6-4n}{1-n}
\end{aligned}
\tag{5.15}
$$

Now consider the following group of independent variables:

$$\hat{x}^{\alpha}\hat{y}^{\beta} = a^{\alpha m+\beta l}(x^{\alpha}y^{\beta}).$$

This group will be invariant if $\alpha = -\frac{\beta}{\gamma}$, where

$$\gamma = \frac{m}{l} = \frac{6-4n}{1-n}. \tag{5.16}$$

Thus we have the following possible similarity variables:

$$\eta(x, y) = k_1 e^{-k_2(yx^{-1/\gamma})^{\beta}} \tag{5.17}$$

$$\eta(x, y) = k_2(yx^{-1/\gamma})^{\beta} \tag{5.18}$$

where k_1, k_2 and β are arbitrary constants.

Next consider the complete group with respect to the stream function $\psi(x, y)$,

$$\hat{x}^{\mu_1}\hat{y}^{\mu_2}\hat{\psi}^{\mu_3} = a^{m\mu_1+l\mu_2+p\mu_3}(x^{\mu_1}y^{\mu_2}\psi^{\mu_3}).$$

This group will be invariant if

$$\mu_3 = -\frac{\mu_2+\gamma\mu_1}{\gamma-1}.$$

Without loss of generality, choose $\mu_2 = 0$, and $\mu_1 = 1$. With this choice we can write the transformation of the stream function $\psi(x, y)$ as follows:

$$\psi(x, y) = \lambda_1 x^{\frac{\gamma-1}{\gamma}} f_1(\eta) \tag{5.19}$$

where λ_1 is an arbitrary constant and $f_1(\eta)$ is a function of η alone.

Similarly, we can write the transformation of the concentration distribution as follows:

$$C(x, y) = \lambda_2 x^{\frac{\gamma-4}{\gamma}} f_2(\eta) \tag{5.20}$$

where λ_2 is an arbitrary constant, and $f_2(\eta)$ is another function of η alone. These general transformation group (5.17), (5.18), (5.19) and (5.20) are found to be valid except for the order of reaction $n = 1, \frac{3}{2}$ and $n = \frac{5}{3}$.

Using relations (5.17), (5.19) and (5.20), the boundary conditions (5.9) and (5.10) imply that

$$\begin{aligned} \eta = 0: \quad & f_1(0) = 0, \\ & f_2(0) = 0, \end{aligned} \tag{5.21}$$

$$\begin{aligned} \eta = k_1: \quad & f_1(k_1) = 0, \\ & f_1'(k_1) = 0, \\ & f_2(k_1) = \frac{1}{\lambda_2}\left\{x^{\frac{4-\gamma}{\gamma}} C_0(x)\right\}. \end{aligned} \tag{5.22}$$

Also using relations (5.18), (5.19) and (5.20), the boundary conditions (5.9) and (5.10) would require that

$$\begin{aligned}\eta = 0: \quad & f_1(0) = 0,\\ & f_1'(0) = 0,\\ & f_2(0) = \frac{1}{\lambda_2}\left\{x^{\frac{4-\gamma}{\gamma}} C_0(x)\right\}. \end{aligned} \tag{5.23}$$

$$\begin{aligned}\eta = \infty: \quad & f_1(\infty) = 0,\\ & f_2(\infty) = 0. \end{aligned} \tag{5.24}$$

5.4.1 Use of the similarity variable $\eta(x, y) = k_1 e^{-k_2(yx^{-1/\gamma})^\beta}$

Thus for $\beta \neq 1$, with the transformations (5.17), (5.19) and (5.20), the following ordinary differential equations result:

The momentum equation is

$$\begin{aligned}
&\lambda_1 k_2 \beta \left(\frac{\gamma-2}{\gamma}\right)\left(-\frac{\ln(\eta/k_1)}{k_2}\right)^{\frac{\beta+1}{\beta}} (\eta f_1')^2\\
&+ \lambda_1(\beta-1)\left(\frac{\gamma-1}{\gamma}\right)\left(-\frac{\ln(\eta/k_1)}{k_2}\right)^{\frac{1}{\beta}} \eta f_1 f_1'\\
&- \lambda_1 k_2 \beta \left(\frac{\gamma-1}{\gamma}\right)\left(-\frac{\ln(\eta/k_1)}{k_2}\right)^{\frac{\beta+1}{\beta}} (\eta f_1 f_1' + \eta^2 f_1 f_1'')\\
= -\nu\Bigg[&(\beta-1)(\beta-2)\eta f_1' - 3k_2\beta(\beta-1)\left(-\frac{\ln(\eta/k_1)}{k_2}\right)(\eta f_1 + \eta^2 f_1)\\
&+ k_2^2\beta^2\left(-\frac{\ln(\eta/k_1)}{k_2}\right)^2 (\eta f_1' + 3\eta^2 f_1'' + \eta^3 f_1''')\Bigg]\\
&+ \frac{g\beta^*\lambda_2}{\lambda_1 k_2 \beta}\left(-\frac{\ln(\eta/k_1)}{k_2}\right)^{\frac{3-\beta}{\beta}} f_2
\end{aligned} \tag{5.25}$$

The mass-diffusion equation is

$$\begin{aligned}
&\lambda_1\lambda_2 k_2 \beta \left(-\frac{\ln(\eta/k_1)}{k_2}\right)^{\frac{\beta-1}{\beta}} \left(\frac{\gamma-1}{\gamma}\eta f_1 f_2' - \frac{\gamma-4}{\gamma}\eta f_1' f_2\right)\\
&= -D\lambda_2\beta k_2\Bigg[(\beta-1)\left(-\frac{\ln(\eta/k_1)}{k_2}\right)^{\frac{\beta-2}{\beta}} \eta f_2'\\
&-k_2\beta\left(-\frac{\ln(\eta/k_1)}{k_2}\right)^{\frac{2\beta-2}{\beta}} (\eta f_2' + \eta^2 f_2'')\Bigg] - k\lambda_2^n f_2^n
\end{aligned} \tag{5.26}$$

For $\beta = 1$, equations (5.25) and (5.26) reduce to the following simple forms:

The momentum equation is

$$f_1' + 3\eta^2 f_1'' + \eta^3 f_1''' - f_2 + \left\{ \frac{\gamma - 2}{\gamma}(\eta f_1')^2 - \frac{\gamma - 1}{\gamma}(\eta f_1 f_1' + \eta^2 f_1 f_1'') \right\} = 0 \tag{5.27}$$

The mass-diffusion equation is

$$\frac{1}{Sc}\left\{ \eta f_2' + \eta^2 f_2'' \right\} + \frac{\gamma - 4}{\gamma}\eta f_1' f_2 - \frac{\gamma - 1}{\gamma}\eta f_1 f_2' - \left(\frac{k\lambda_2^{n-3/2}}{\sqrt{g\beta^*}} \right) f_2^n = 0. \tag{5.28}$$

5.4.2 Use of the similarity variable $\eta(x, y) = k_2(yx^{-1/\gamma})^\beta$

In this case again, for $\beta \neq 1$, with the transformation (5.18), (5.19) and (5.20), we obtain the following ordinary differential equations:

The momentum equation is

$$\begin{aligned} \nu E\beta^2 f_1''' &= 2\beta(2 - n)\eta^{\frac{1-\beta}{\beta}} f_1'^2 - (\beta - 1)(5 - 3n)\eta^{\frac{1-2\beta}{\beta}} f_1 f_1' \\ &\quad - \beta(5 - 3n)\eta^{\frac{1-\beta}{\beta}} f_1 f_1'' \\ &\quad - \nu E\left((\beta - 1)(\beta - 2)\eta^{-2} f_1' + 3\beta(\beta - 1)\eta^{-1} f_1'' \right) \\ &\quad - g\beta^* M\eta^{\frac{3(1-\beta)}{\beta}} f_2 \end{aligned} \tag{5.29}$$

The mass-diffusion equation is

$$\begin{aligned} DE\beta f_2'' &= \eta^{\frac{1-\beta}{\beta}} \left\{ 2f_1' f_2 + (3n - 5) f_1 f_2' \right\} \\ &\quad - DE(\beta - 1)\eta^{-1} f_2' + kG\eta^{2\frac{1-\beta}{\beta}} f_2^n \end{aligned} \tag{5.30}$$

where

$$E = \frac{k_2^{1/\beta}(6 - 4n)}{\lambda_1}$$

$$M = \frac{\lambda_2(6 - 4n)}{\beta\lambda_1^2 k_2^{2/\beta}}$$

$$G = \frac{\lambda_2^{n-1}(6 - 4n)}{\beta\lambda_1 k_2^{1/\beta}}$$

For $\beta = 1$, equations (5.29) and (5.30) reduce to the simple forms:

$$f''' + f_2 + \frac{\gamma - 1}{\gamma} f_1 f_1'' - \frac{\gamma - 2}{\gamma} f_1'^2 = 0, \tag{5.31}$$

$$\frac{1}{Sc} f_2'' + \frac{\gamma - 1}{\gamma} f_1 f_2' - \frac{\gamma - 4}{\gamma} f_1' f_2 - \frac{k\lambda_2^{n-1}}{\nu k_2^2} f_2^n = 0. \tag{5.32}$$

The boundary conditions for (5.29) and (5.30) or for (5.31) and (5.32) are given by (5.23) and (5.24).

Remark

Examination of the boundary conditions (5.22) and (5.23) reveals that similarity analysis based on the transformation group (5.11) is not meaningful when (a) $C_0(x) =$ constant, i.e., uniform concentration along the plate; (b) $C_0(x) = F(x)$, i.e., the most general condition where F is an arbitrary function of x alone. However, a tractable problem can be obtained provided that

$$C_0(x) = Nx^{\frac{\gamma-4}{\gamma}} \tag{5.33}$$

It appears from this analysis that a formal simplification of the problem using similarity analysis with the simplified transformation group (5.11), that is a precise definition of the problem in terms of ordinary differential equations, is possible only if the initial concentration is described along the plate according to (5.33). This would be useful in the case where the prescribed distribution approximates a simple power-law where the form depends on the particular order of reaction of the system. In practice, depending on the particular chemical constituents chosen, it may or may not be possible to achieve and maintain in steady-state a concentration distribution along the plate in accordance with (5.33), such as to permit useful application of this approach. The mathematical treatment is valid, however, and some theoretical inferences may be of considerable academic interest.

5.5 Natural convection flow along a vertical plate

A study of laminar natural convection flow over a semi-infinite vertical plate at constant species concentration is examined. The plate is maintained at a given concentration of some chemical species while convection is induced by diffusion into and chemical reaction with the ambient fluid. In the absence of chemical reaction, a similarity transform is possible. When chemical reaction occurs, perturbation expansions about an additional similarity variable dependent on reaction rate must be employed. Two fundamental parameters of the problem are the Schmidt number, Sc, and the reaction order, n. Results are presented for the Schmidt number ranging from 0.01 to 10000 and reaction order up to 5. In the presence of a chemical reaction, the diffusion and velocity domains expand out from the plate. This results in a larger, less distinct convection layer.

Natural Convection flow near a semi-infinite vertical plate has been studied extensively in the existing literature [5–7, 9, 15]. These studies have included the effect of temperature gradients on the flow of fluid medium. Situations where the plate is held at uniform temperature, as well as with a step change in wall temperature, have been studied [13, 16]. Yang et al. [21] extended the study to include a nonisothermal vertical plate immersed in a temperature stratified medium. Kulkarni et al. [11] considered natural convection flow over an isothermal vertical wall immersed in a thermally stratified medium. Angirasa et al. [2] studied heat and mass transfer by natural convection adjacent to vertical surfaces situated in fluid-saturated porous media. They give special attention to opposing buoyancy effects of the same order and unequal thermal and species diffusion coefficients. Their numerical results support the validity of the boundary layer analysis for high Rayleigh number aiding flows and for opposing flows when one of the buoyant forces overpowers the other.

Levich [14] reported on a problem of steady-state natural convection induced by chemical diffusion. In this formulation, the plate was held at zero concentration of a chemical species, A, containing catalytic substances. Upon contact with a fluid solution, a heterogeneous chemical reaction occurs at the plate which in the presence of a gravitational field led to natural convection flow near the plate. Kostin and Gray [10] studied the problem in which a homoegeneous chemical reaction takes place in a vertical, adiabatic, steady-state flow reactor. They observed that the release of heat by the chemical reaction(exothermic reaction) sets up density gradients which cause natural convection flow to be superimposed on the laminar up-flow. In their analysis, they observed that the heat generated by the chemical reaction caused temperature gradients which produced density gradients which in turn caused natural convection flow to be superimposed on the laminar up-flow. Gebhart and Pera [7] investigated natural convection flows caused by the simultaneous diffusion of thermal energy and of chemical species. They assumed small species concentration levels and showed that the Boussinesq approximations led to similarity solutions similar in form to those found for single buoyancy mechanism flows.

Meadley and Rahman [15] considered the effects of chemical diffusion and reaction from a vertical plate and presented analytical and numerical solutions for the ranges $10^{-2} \leq Sc \leq 10^{4}$ and $0 \leq n \leq 5$. They found that the presence of chemical reaction expanded the diffusion and velocity domains out from the plate resulting in a larger, less distinct convection layer. Additional work in this direction has not been actively pursued in recent years. The present work examines natural convection flow in the presence of both chemical reaction and diffusion and the resulting flow patterns. Of particular interest is the observed flow reversal for higher Schmidt numbers.

5.6 Mathematical formulation

A vertical plate is composed of a chemical species maintained at a given concentration and immersed in a fluid. The ambient fluid is similarly maintained at a constant but distinct concentration away from the plate. The species on the plate is first transferred from the plate to the adjacent fluid by diffusion after which a homogeneous isothermal irreversible chemical reaction of order n is assumed to occur in the fluid between the constituents of the plate and fluid. We have made the following assumptions:

- the fluid is Newtonian
- two-dimensional laminar steady flow is considered
- the physical properties associated with the problem, such as viscosity, Diffusivity, etc. are assumed constant
- the Boussinesq approximation is taken into consideration for buoyancy effects, which implies small variations in the properties which lead to buoyancy forces
- the reaction number is small
- the static pressure gradients arising from the convection currents are neglected.

A change in chemical composition of the fluid near the surface of the plate produces a lighter fluid which rises due to the induced buoyancy forces. The result is an upward movement of fluid particles near the plate surface. If a heavier fluid is produced, the reverse effect is observed but the mathematical formulation is still the same. When a chemical reaction occurs in the bulk of the fluid, the diffusing species may be depleted. The equations governing the flow are derived from conservation of mass (fluid and species) and momentum. They can be written as [15]:

$$\frac{\partial U}{\partial X} + \frac{\partial V}{\partial Y} = 0 \tag{5.34}$$

$$U\frac{\partial U}{\partial X} + V\frac{\partial U}{\partial Y} = g\beta^*(C - C_\infty) + \nu\frac{\partial^2 U}{\partial Y^2} \tag{5.35}$$

$$U\frac{\partial C}{\partial X} + V\frac{\partial C}{\partial Y} = \dot{C}''' + D\frac{\partial^2 C}{\partial Y^2} \tag{5.36}$$

where X is the coordinate chosen vertically upwards along the plane, Y is the horizontal coordinate perpendicular to X, U is the velocity component along X, V the velocity component along Y, C is the species concentration, C_∞ the species concentration at infinity, D is the diffusion coefficient, ν the dynamic viscosity, g the acceleration due to gravity, β^* is the volumetric expansion of concentration and $\dot{C}'''$ is the reaction rate term. The reaction rate term $\dot{C}'''$ represents the reaction kinetics of the system whose overall reaction is described by the power-law model. This term takes the form

$$\dot{C}''' = -k(C - C_\infty)^n$$

where k is the reaction rate constant and n the order of the reaction (see Aris [4] for details). To completely describe the problem, suitable boundary conditions are required. These are given by

$$\left.\begin{array}{llll} \text{at} \quad Y = 0, & U = V = 0, & \forall X & \text{(no slip condition)} \\ \text{at} \quad Y \to \infty, & U = V = 0, & \forall X & \text{(uniformity at } \infty) \\ \text{and} & (C - C_\infty) \to 0, & \forall X & \text{(uniformity at } \infty) \end{array}\right\} \tag{5.37}$$

In addition, at the plate surface $Y = 0$,

$$C = C_0(X) \tag{5.38}$$

The nature of the function $C_0(X)$ along the plate length may be subject to severe limitations which arise from the chemical kinetics involved in setting up steady-state conditions for a given species and ambient fluid. For the formulation of the mathematical model, the concentration of the species at the plate surface is maintained (by some external means) at uniform concentration C_0. Using the definition of the two-dimensional stream function $\psi(X, Y)$ as $U = \psi_Y$ and $V = -\psi_X$, the governing equations (5.34)–(5.36) can be simplified as follows:

$$\psi_Y \psi_{YX} - \psi_X \psi_{YY} = \nu \psi_{YYY} + g\beta^*(C - C_\infty) \tag{5.39}$$

$$\psi_Y C_X - \psi_X C_Y = DC_{YY} - k(C - C_\infty)^n \tag{5.40}$$

The boundary conditions (5.37)) and (5.38) are:

$$Y = 0: \quad \left.\begin{array}{c} \psi_Y = \psi_X = 0 \\ C = C_0(X) \end{array}\right\} \tag{5.41}$$

$$Y \to \infty: \quad \left.\begin{array}{c} \psi_Y = \psi_X = 0 \\ C - C_\infty = 0 \end{array}\right\} \tag{5.42}$$

The following similarity transformations are then introduced into the above equations:

$$c = \frac{C - C_\infty}{C_0 - C_\infty} \tag{5.43}$$

$$\eta(X, Y) = \frac{Y}{X}\left\{\frac{Gr_x}{4}\right\}^{1/4} \tag{5.44}$$

$$\psi(X, Y) = 4\nu\left\{\frac{Gr_x}{4}\right\}^{1/4} f(\eta) \tag{5.45}$$

where

$$Gr_x = \frac{g\beta^* X^3(C_0 - C_\infty)}{\nu^2}. \tag{5.46}$$

Here c is the dimensionless species concentration, η is the similarity variable, $f(\eta)$ is the dimensionless stream function and Gr_x is the Grashof number. Grashof number represents the vigour of the buoyancy force exerted on the fluid. The transformations produce the ordinary differential equations shown below [8].

$$f''' + c + 3ff'' - 2f'^2 = 0 \tag{5.47}$$

$$\frac{c''}{Sc} + 3fc' - \epsilon(X)c^n = 0 \tag{5.48}$$

where

$$\epsilon(X) = \frac{2k(C_0 - C_\infty)^{n-3/2}}{\sqrt{g\beta^*}} X^{1/2}$$

and $Sc = \frac{\nu}{D}$ is the Schmidt number.

In the absence of chemical reaction, the similarity analysis is complete. However, in the presence of a chemical reaction, the local reaction-rate number, $\epsilon(X)$, may be considered arbitrarily small depending on the " smallness" of the reaction. Most chemical species with slow reaction rate have small reaction numbers. As an example, chemicals such as sodium chloride or potassium chloride with dilute acids, and sugar with dilute acids can be categorised in this class of species. Recall that the basic perturbation expansion is [20]

$$c(\eta, \epsilon) = c_0(\eta) + \epsilon(X)c_1(\eta) + \epsilon^2(X)c_2(\eta) + \dots \tag{5.49}$$

Suppose $|\frac{c_2}{c_1}| \leq K$, where K is the upper bound which is known numerically for all η in the calculation. Assume that $\lim_{\eta\to\infty} |\frac{c_2}{c_1}| = 0$ and that the perturbation expansion (5.49) is a convergent series. If we let

$$\left|\frac{\epsilon^2(X)c_2(\eta)}{\epsilon(X)c_1(\eta)}\right| \leq 0.1\%, \quad \text{say, for} \quad 0 < \eta < \infty$$

so that the error in using the truncated solution

$$c = c_0(\eta) + \epsilon(X)c_1(\eta) \tag{5.50}$$

is of order 0.1% or $|\epsilon(X)| < 0.001|\frac{c_1(\eta)}{c_2(\eta)}|$. Since $|\frac{c_1}{c_2}|$ has the lower bound $\frac{1}{K}$, we therefore require that

$$|\epsilon(X)| \leq \frac{0.001}{K} \tag{5.51}$$

in order to ensure the desired accuracy. Let $\frac{0.001}{K} = \xi$, so that $0 < |\epsilon(X)| \leq \xi$ and $\epsilon(X)$ must be of order ξ. But the requirement

$$|\epsilon(X)| = \frac{2k(C_0 - C_\infty)^{n-3/2}}{(g\beta^*)^{1/2}} |X^{1/2}| \leq \xi$$

implies that $0 < x \leq \xi^2$, where $x = \frac{4k^2(C_0-C_\infty)^{2n-3}}{g\beta^*} X$ would appear to be the approximate non-dimensional plate coordinate. Thus, a solution may be possible by perturbation expansion about $\epsilon(X)$ of the order ξ - provided attention is confined to the region downstream of leading edge given by $0 < x \leq \xi^2$. In this chapter, we have used the value of $\epsilon(X) = 0.01$. With these additional observations, the similarity transformations take the form

$$c(X, Y) = c(\eta, \epsilon) \tag{5.52}$$

$$\psi(X, Y) = 4\upsilon \left\{\frac{Gr_x}{4}\right\}^{1/4} f(\eta, \epsilon) \tag{5.53}$$

where

$$\eta(X,Y) = \frac{Y}{X}\left\{\frac{Gr_x}{4}\right\}^{1/4} \tag{5.54}$$

$$\epsilon(X) = \frac{2k(C_0 - C_\infty)^{n-3/2}}{\sqrt{g\beta^*}}X^{1/2} \tag{5.55}$$

These transformations can be combined with a regular perturbation scheme about ϵ as shown below. Since we are considering a regular perturbation problem, there is no need for the matching of layers or for multiple scales.

$$f(\eta,\epsilon) = f_0(\eta) + \epsilon f_1(\eta) + \epsilon^2 f_2(\eta) + \ldots \tag{5.56}$$

$$c(\eta,\epsilon) = c_0(\eta) + \epsilon c_1(\eta) + \epsilon^2 c_2(\eta) + \ldots \tag{5.57}$$

The transformations combined with the perturbation expansions about ϵ lead to the following sets of ordinary differential equations up to first order in ϵ:

- Zeroth-order approximation

$$f'''_0 + c_0 + 3f_0 f''_0 - 2f'^2_0 = 0 \tag{5.58}$$

$$c''_0 + 3Sc f_0 c'_0 = 0 \tag{5.59}$$

- First-order approximation

$$f_1''' + c_1 + 5f_1 f''_0 + 3f_0 f_0'^2 = 0 \tag{5.60}$$

$$c_1'' + Sc(5f_1 c_0' + 3f_0 c_1' - 2c_1 f_0' - c_0{}^n) = 0 \tag{5.61}$$

These equations parallel to those found in Gebhart et al. [8] and Merkin and Mahmood [17]. Note that for the first order equations, there are terms from the zeroth-order equations coupled with first order terms. For each subsequent order (second, third, etc.) the equations will contain terms from the previous orders coupled with the present order terms as seen in this case. The boundary conditions for these approximations may be written as

$$\text{at} \quad \eta = 0: \quad \left.\begin{aligned} f_r(0) &= 0 \\ f_r'(0) &= 0 \\ c_0(0) &= 1 \\ c_{r+1}(0) &= 0 \end{aligned}\right\} \tag{5.62}$$

when $\eta \to \infty$, $f_r'(\infty) = c_r(\infty) = 0$ where $r = 0, 1$.

Equations (5.58) to (5.61) have been integrated numerically using the classical fourth-order Runge-Kutta method in combination with the shooting technique for determining correct initial conditions. These initial conditions at the surface are needed so that the asymptotic boundary conditions may be satisfied. The results are presented for the case where no chemical reaction occurs and when chemical reactions of order up to $n = 5$ occur.

It must be noted that it is possible to obtain equations with fractional reaction orders, i.e. $n = \frac{1}{2}$, $n = \frac{3}{2}$, etc. This further analysis has been performed by the authors and is to be presented for publication in a later work. In the fractional reaction order case, a similarity analysis is used with a group theoretic approach to obtain the relevant equations for each reaction order. The equations obtained through this analysis can be solved using the same numerical technique described in the present work. For the benefit of the reader a general numerical scheme has been presented in the following section.

5.7 Method of numerical solution

These approximations have been integrated numerically for various Schmidt numbers and reaction orders using a Runge-Kutta integration scheme to correct for assumed starting values of the initial conditions at the surface. The general forms of the equations to be treated are:

$$f'''(\eta) = F(f'', f', g', f, g, \eta) \tag{5.63}$$

$$g''(\eta) = G(f'', f', g', f, g, \eta) \tag{5.64}$$

with the initial and asymptotic boundary conditions (5.62). With the two asymptotic boundary conditions, it was necessary to assume starting values for the two missing conditions at $\eta = 0$ that were required. Let

$$A = f''(0) \tag{5.65}$$

$$B = g'(0) \tag{5.66}$$

to fulfill the requirement that

$$\lim_{\eta \to \infty} f'(A, B, \eta) = f'_\infty(A, B) = 0 \tag{5.67}$$

$$\lim_{\eta \to \infty} g(A, B, \eta) = g_\infty(A, B) = 0 \tag{5.68}$$

If it is assumed that A_1 and B_1 are trial values of A and B such that

$$A = A_1 + h$$
$$B = B_1 + k$$

where h and k are small and thus by (5.67) and (5.68) we have

$$f'_\infty(A_1 + h, B_1 + k) = 0 \tag{5.69}$$

$$g_\infty(A_1 + h, B_1 + k) = 0 \tag{5.70}$$

In addition, to satisfying the asymptotic boundary conditions, it was assumed that the gradients of (5.69) and (5.70) were zero at infinity. This leads to

$$f''_\infty(A_1 + h, B_1 + k) = 0 \tag{5.71}$$

$$g'_\infty(A_1 + h, B_1 + k) = 0 \tag{5.72}$$

Taylor's expansions for small h and k were then applied to equations (5.69) to (5.72). In matrix form, the problem may be expressed as:

$$\begin{pmatrix} \dfrac{\partial f'_\infty}{\partial A} & \dfrac{\partial f'_\infty}{\partial B} \\ \dfrac{\partial g_\infty}{\partial A} & \dfrac{\partial g_\infty}{\partial B} \\ \dfrac{\partial f''_\infty}{\partial A} & \dfrac{\partial f''_\infty}{\partial B} \\ \dfrac{\partial g'_\infty}{\partial A} & \dfrac{\partial g'_\infty}{\partial B} \end{pmatrix} \begin{pmatrix} h \\ k \end{pmatrix} = - \begin{pmatrix} f'_\infty(A, B) \\ g_\infty(A, B) \\ f''_\infty(A, B) \\ g'_\infty(A, B) \end{pmatrix} \tag{5.73}$$

The application of the least squares method yields the least square error shown below. The partial derivatives appearing in the solutions of h and k can be obtained by integrating the perturbed differential equations with their appropriate initial conditions. From equations (5.63) and (5.64) we obtain the perturbed differential equations for the A-derivatives with the initial conditions as shown in the following:

$$\frac{\partial f'''}{\partial A} = \frac{\partial F}{\partial f''}\frac{\partial f''}{\partial A} + \frac{\partial F}{\partial f'}\frac{\partial f'}{\partial A} + \frac{\partial F}{\partial g'}\frac{\partial g'}{\partial A} + \frac{\partial F}{\partial f}\frac{\partial f}{\partial A} + \frac{\partial F}{\partial g}\frac{\partial g}{\partial A} \tag{5.74}$$

$$\frac{\partial g''}{\partial A} = \frac{\partial G}{\partial f''}\frac{\partial f''}{\partial A} + \frac{\partial G}{\partial f'}\frac{\partial f'}{\partial A} + \frac{\partial G}{\partial g'}\frac{\partial g'}{\partial A} + \frac{\partial G}{\partial f}\frac{\partial f}{\partial A} + \frac{\partial G}{\partial g}\frac{\partial g}{\partial A} \tag{5.75}$$

$$\left.\begin{aligned} \eta = 0: \quad \frac{\partial f}{\partial A} = \frac{\partial f'}{\partial A} = \frac{\partial g'}{\partial A} = \frac{\partial g}{\partial A} = 0 \\ \frac{\partial f''}{\partial A} = 1 \end{aligned}\right\} \tag{5.76}$$

Perturbed differential equations for the B-derivatives with appropriate initial conditions were obtained from equations (5.63) and (5.64). To correct the trial values of A_1 and B_1, the original equations (5.63) to (5.66) with the perturbed equations for A and B with their appropriate initial conditions were integrated simultaneously up to a certain suitable point. At this point, the trial values of A_1 and B_1 were corrected to refine the solution. After two or three iterations at the same point where the least-square error appeared to be steady, the integration range was extended and the process repeated up to the extended point. This iterative process was continued until the desired solution accuracy was obtained.

5.8 Numerical results

The results obtained by the numerical solution procedure introduced in the preceding section are fairly consistent with that in the literature. The convergence of the procedure is fairly rapid and is found to depend upon the choice of step-size. It is observed that for larger step-sizes, convergence is slower than when smaller step-sizes are used. This seems to indicate some stiffness in the solution. The graphical output of the results reveals that there is a great degree of variability in the solution close to the plate surface. The results obtained here are similar to those obtained by Kostin and Gray [10] although the geometries are different. They found that the flow affects the concentration and temperature distributions which in turn can affect the flow. Their analysis included a homogeneous chemical reaction which was exothermic and thus caused temperature gradients which in turn produced density gradients. The present analysis involves a constant temperature formulation. It is observed that the velocity profile in the present analysis is also parabolic near the wall. However, unlike the procedure in their case, rapid convergence is obtained with the correct choice of step-size.

The results are comparable to those found in Gebhart and Pera [7]. For the zeroth order perturbation solution, the concentration profiles reveal that increasing the Schmidt number reduces the species diffusion layer (Fig. 5.1). At the same time, the velocity level is reduced as seen in the velocity profiles (Fig. 5.2). These results confirm Gebhart and Pera's observations that aiding buoyancy effects decrease the diffusion boundary region and at the same time increases the velocity level (and its extent) as the Schmidt number decreases.

In the case of first order perturbation solution, numerical results are plotted in Figs. 5.3 and 5.4. The velocity and concentration profiles for $n = 1, 2, 3, 5$ and $Sc = 10$ are plotted in Figs. 5.3 and 5.4. It is seen from these figures that the extent of these profiles increases as the order of reaction rate increases for fixed Schmidt number. This result agrees with the physical situation.

From Fig 5.4, as the chemical species on the plate reacts with the fluid and rises (due to buoyancy changes), dense fluid is brought into contact with the plate. What then occurs is a downward movement of fluid above to replace the less dense, rising fluid. Even though the less dense fluid is lighter, it appears that in the particular

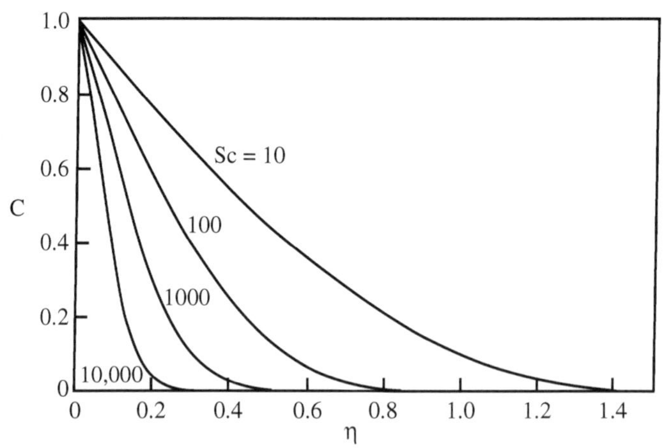

Figure 5.3: Dimensionless concentration profiles without reaction (zeroth-order approximation) for $Sc = 10, 100, 1000, 10000$ (from [18]).

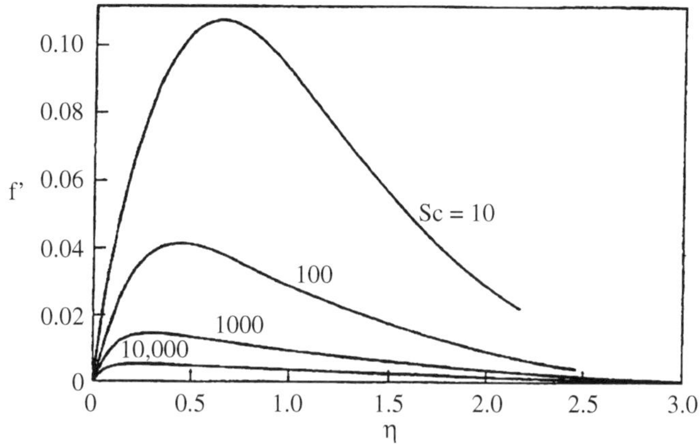

Figure 5.4: Dimensionless velocity profiles without reaction (zeroth-order approximation) for $Sc = 10, 100, 1000, 10,000$ (from [18]).

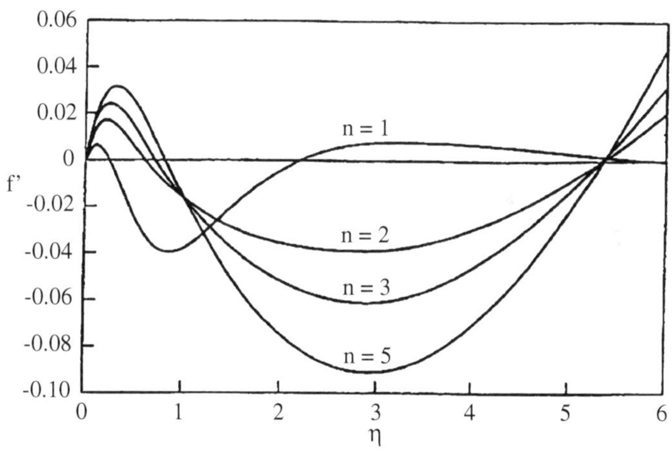

Figure 5.5: Dimensionless velocity profiles with reaction (up to first-order approximation) for $Sc = 10$ and $n = 1, 2, 3, 5$ (from [18]).

cases depicted ($Sc = 10$), the more dense fluid from above displaces the less dense fluid by moving downward. In addition to this movement, fluid from the region near the plate, which is also less dense than the rising fluid, moves to displace the lighter fluid resulting in the upward movement of lighter fluid in the region near the plate.

Fig. 5.5 and Fig. 5.6 displayed the dimensionless velocity profiles and the dimensionless concentration profiles, respectively, for various parameters of the problem studied here. The results showed fair agreement with the available experimental data.

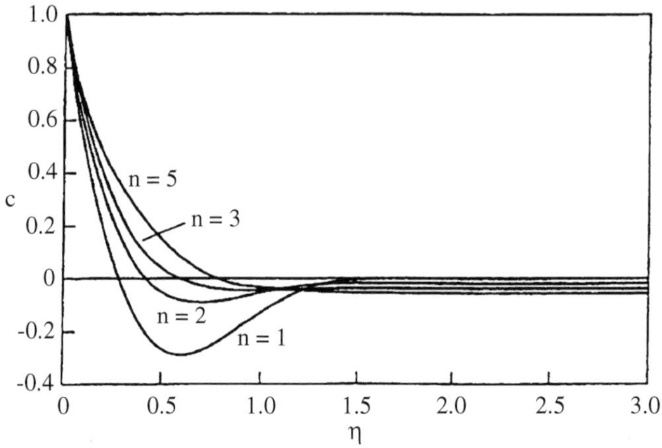

Figure 5.6: Dimensionless concentration profiles with reaction (up to first-order approximation) for $Sc = 10$ and $n = 1, 2, 3, 5$ (from [18]).

For further information about this problem, the reader is referred to the work of Ames, W.F. [1], Angirasa, D. & Srinivasan, J. [3], Lamb, H. [12], and Reynolds, O. [19] as listed in the reference section.

5.9 Exercises

1. Consider the heat conduction in a semi-infinite medium. In one-dimensional case the partial differential equation is given by $u_t = \alpha u_{xx}$ subject to the boundary conditions $x = 0 : u(x, 0) = u_0$, and $x \to \infty : u(x, t) = 0$; and the initial condition $t = 0 : u(x, 0) = 0, x > 0$. Using similarity method find the solution of problem. [Hint: It can be shown that the similarity variable is $\eta(x, t) = \frac{x}{2\sqrt{\alpha t}}$. Use the dependent variable as $u = u_0 f(\eta)$ and then show that this initial boundary problem can ce written as $f'' + 2\eta f' = 0$; $\eta(0) : f = 1$; $\eta \to \infty$; $f = 0$. The solution is $f = \frac{u}{u_0} = erfc(\eta) = \frac{2}{\sqrt{\pi}} \int_\eta^\infty exp(-z^2)dz$.
2. A semi-infinite flat plate is immersed in a steady uniform stream of an incompressible fluid with a viscosity ν. The flow behaviour about the flat plate is considered to be boundary-layer flow in two dimensions. The mathematical theory of boundary layer flow was first developed by Ludwig Prandtl in 1904. The mathematical equations are given as: Momentum equation is $uu_x + vu_y = \nu u_{yy}$; Continuity equation is: $u_x + v_y = 0$. The boundary conditions are given by $u = v = 0$; at $y = 0$; $u = u_0$; $y \to \infty$. Determine the solution of this boundary value problem. [$\eta = y\sqrt{u_0/2\nu x}$ and $\psi(x, y) = u_0\sqrt{2\nu x/u_0} f(\eta)$, and then show that $f''' + ff'' = 0$.]
3. In cylindrical polar coordinates, the vorticity of a fluid is related to the velocity of the fluid in the following manner, $\omega = \frac{\partial v}{\partial r} + \frac{v}{r}$, where v is the velocity and ω is the vorticity. The vorticity ω satisfies also the equation of motion

$\frac{\partial \omega}{\partial t} = \nu\{\frac{\partial^2 \omega}{\partial r^2} + \frac{1}{r}\frac{\partial \omega}{\partial r}\}$. Find a similarity solution in the form of $\omega = f(\eta)g(t)$, where $\eta = r/2\sqrt{\nu t}$ and verify the velocity distribution obtained by this method is of the form $v = \frac{K}{2\pi r}\{1 - \exp(-\frac{r^2}{4\nu t})\}$, where K is the strength of the vortex line.

4. Consider the following boundary-layer equations $u_x + v_y = 0$; $uu_x + vu_y = \nu u_{yy}$. Using the similarity variables $\psi = u_0 x(\frac{u_0 x}{\nu})^m f(\eta)$; $\eta = \frac{y}{x}(\frac{u_0 x}{\nu})^n$, show that $(\frac{u_0 x}{\nu})^{-m+n-1} f''' + (m+1)ff'' - (m+n)f'^2 = 0$.
5. In Exercise 4, if $n = m + 1$, then the above equation can be written $f''' + (m+1)ff'' - (2m+1)f'^2 = 0$. Now assume $m - \frac{2}{3}]$, show that the solution of this resultant equation may be written as $f(\eta) = C\tanh(\frac{C\eta}{6})$, where C is a constant.

References

[1] Ames, W.F., *Nonlinear Partial Differential Equations in Engineering*, Academic Press: New York, 1965.

[2] Angirasa, D., Peterson, G.P. & Pop, I., Combined heat and mass transfer by natural convection with opposing buoyancy effects in a fluid saturated porous medium. *Int. J. Heat Mass Transfer*, **40(12)**, pp. 2755–2773, 1997.

[3] Angirasa, D. & Srinivasan, J., Natural convection flows due to the combined buoyancy of heat and mass diffusion in a thermally stratified medium. *J. Heat Transfer*, **111**, pp. 657–663, 1989.

[4] Aris, R., *Introduction to the Analysis of Chemical Reactors*, Prentice-Hall, Inc.: New Jersey, 1965.

[5] Bejan, A. & Lage, J.L., The Prandtl number effect on the transition in natural convection along a vertical surface. *J. Heat Transfer*, **112**, pp. 787–790, 1990.

[6] Cheesewright, R., Turbulent natural convection from a vertical plane surface. *Trans. Am. Soc. Mech. Engrs.*, **90(C)**, pp. 1–8, 1968.

[7] Gebhart, B. & Pera, L., The nature of vertical natural convection flows resulting from the combined buoyancy effects of thermal and mass diffusion. *Int. J. Heat Mass Transfer*, **14**, pp. 2025–2049, 1971.

[8] Gebhart, B., Jaluria, Y., Mahajan, R.L. & Sammakia, B., *Bouyancy-induced Flows and Transport*, Hemisphere Publishing Corporation: New York, 1988.

[9] Henkes, R.A.W.M. & Hoogendoorn, C.J., Laminar natural convection boundary-layer flow along a heated vertical plate in a stratified environment. *Int. J. Heat Mass Transfer*, **32**, pp. 147–155, 1989.

[10] Kostin, M.D. & Gray, W.G., Velocity, temperature and concentration profiles in a vertical flow reactor. *Chemical Eng. J.*, **30**, pp. 931–936, 1975.

[11] Kulkarni, A.K., Jacobs, H.R. & Hwang, J.J., Similarity solution for natural convection flow over an isothermal vertical wall immersed in thermally stratified medium. *Int. J. Heat Mass Transfer*, **30**, pp. 691–698, 1987.

[12] Lamb, H., Hydrodynamics, 6th edn, Dover: New York, 1945.

[13] Lee, S. & Yovanovich, M.M., Laminar natural convection from a vertical plate with a step change in wall temperature. *J. Heat Transfer*, **113**, pp. 501–504, 1991.

[14] Levich, V.G., *Physico-chemical Hydrodynamics*, Prentice-Hall, Englewood Cliffs: New Jersey, 1962.

[15] Meadley, C.K. & Rahman, M., Laminar natural convection caused by chemical diffusion and reaction from a vertical plane surface. *Can. J. Chem. Eng.*, **52(5)**, pp. 552–557, 1974.

[16] Merkin, J.H., A note on the similarity solutions for free convection on a vertical plate. *J. Eng. Math.*, **19**, pp. 189–201, 1985.

[17] Merkin, J.H. & Mahmood, T., On the free convection boundary layer on a vertical plate with prescribed surface heat flux. *J. Eng. Math.*, **24**, pp. 95–106, 1990.

[18] Rahman, M., *Steady Natural Convection Flow over a Semi-Infinite Vertical Plate Induced by Diffusion and Chemical Reaction*, PhD Thesis, University of Windsor, Windsor, Ontario, Canada, 1973.

[19] Reynolds, O., An experimental investigation of the circumstances which determine whether the motion of water shall be direct or sinuous and the law of resistance in parallel channels. *Phil. Trans. A*, **174**, pp. 935–982, 1883. Also, *Scientific Papers*, **2**, pp. 51–105, 1883.

[20] Van Dyke, M., *Perturbation Methods in Fluid Mechanics*, Academic Press: New York, 1960.

[21] Yang, K.T., Novotny, J.L. & Cheng, Y.S., Laminar free convection from a nonisothermal plate immersed in a temperature stratified medium. *Int. J. Heat Mass Transfer*, **15**, pp. 1097–1109, 1964.

CHAPTER 6

Turbulence

Professor Jyotiprasad Medhi

Professor Jyotiprasad Medhi was the Professor and Head of the Department of Statistics, Gauhati University, Gauhati, India. He earned his MSc degree in pure mathematics from the University of Calcutta, India, and another MSc in Statistics (by thesis) from the University of Manchester, UK, and his doctorate in Statistics (Stochastic Processes) from the University of Paris, France in 1956. Since 1949 he has been a Faculty member of the Post-Graduate Department of Mathematics and Statistics, Gauhati University. He was a Visiting Professor at the University of Montreal, Canada in 1967-1968, and was also on a visiting assignment in the University of

Toronto, Canada during Fall 1982. In 1979, he visited France and UK on a study-cum-lecture tour and for exchange ideas with academics in those countries. He has published extensively in some reputed journals and has written a number of books in Statistics. Professor Medhi was also the Director of Population Research Centre set up by the Ministry of Health and Family Welfare, Government of India in his department. He has been a member of the Institute of Mathematical Statistics, USA since 1963 and was President, Section of Statistics, Indian Science Congress Association 1978–1979.

This chapter Fluid Mechanics Memoir *is dedicated to my revered Sir, Dr Jyotiprasad Medhi, Professor Emeritus, Department of Statistics, Gauhati University, Guwahati, India whose enormous inspirations, encouragements and loving care to me augment my love for higher education in mathematics. I received my fundamental background in mathematics and statistics since my school days in Assam, India and because of his sound mathematical advice in my post-secondary education, I am here in Canada today.*

6.1 Introduction

This chapter contains the study of two important areas of fluid motions. We first start with the preliminaries of turbulent flow and then its application is demonstrated in nonlinear wave–wave interactions. The Boltzmann integral equation is modified to obtain the nonlinear energy spectrum.

The study of fluid flow may be conveniently manifested considering three main classes, namely, steady laminar flow, unsteady laminar flow and turbulent flow. The essential feature of the steady laminar flow is that if some property of the fluid flow is measured, as for example the pressure, we can assert that the value at any point will not vary with time in a given experiment, nor will it vary from experiment to experiment, in other words it is both determinate and non-fluctuating. Then this kind of fluid flow is classified as the steady laminar flow. Unsteady laminar flow differs from the steady laminar flow in that although the pressure measured at any instant at corresponding points will not differ from experiment to experiment, they will vary with time. In other words, the pressure is determinate but fluctuating. Turbulent flow, on the other hand, is characterized by the fact that the pressure measured at a point is random in nature. If the same experiment is performed several times under apparently idealized conditions, then the pressure will not be the same in different experiments, but will fluctuate randomly. In other words, it is not determinate in any simple sense.

Osborne Reynolds was the first to illustrate these differences of fluid flow structures by performing classical experiments. In these experiments, water was allowed to flow slowly through various circular pipes of diameter of up to 2 inches, the mouth of the pipes being bell-shaped to ensure a smooth entry flow. At the entrance of the pipe, water with coloring matters (dye) was introduced, and its subsequent behaviour was examined. It was observed that at sufficiently low velocity the dye was seen to be drawn out in a single straight filament right through the pipe, this behaviour indicates the steady laminar flow. When the water in the supply-tank was made to oscillate, this causes the filament to fluctuate in a sinusoidal manner but well-defined form, which was characterized as an unsteady laminar flow. Returning now to the case of steady laminar flow, when the velocity is sufficiently increased then, no matter how carefully this is done, and no matter how smooth the entry flow is, a point is ultimately reached at which the straight filament breaks down into series of eddies, which become progressively more and more unstable and random in nature if the velocity is further increased. Such a flow is called turbulent flow, the term being due to Reynolds. It was discovered by Reynolds that the parameters determining the onset of this turbulent flow is $\overline{U}d/\nu$, where $\overline{U}$ is the mean velocity through the pipe, d is the diameter of the pipe and ν is the kinematic viscosity of the fluid. This number, namely, $R = \frac{\overline{U}d}{\nu}$ is a dimensionless number and subsequently became known as the Reynolds number. This is a highly significant number to determine whether a fluid flow is laminar or turbulent.

This change-over from laminar flow to turbulent flow is known as the transition to turbulence, and the Reynolds number at which transition occurs is called the critical Reynolds number. In the experiment performed by Reynolds the critical Reynolds number was approximately 13,000. It has subsequently found, however, laminar flow can still be achieved at much higher Reynolds numbers, provided that sufficient care is taken to ensure a perfect smooth entry flow.

6.2 The mechanism of transition to turbulence

To determine the nature of transition to turbulence is a very complex situation. Most of the mathematical investigations have considered the problem of stability and transition to turbulence of the so-called two-dimensional parallel flows, where the basic flow is parallel in nature, being of the form

$$u = U(y), \qquad v = w = 0 \tag{6.1}$$

in two-dimensions or

$$u = U(r), \qquad v = w = 0 \tag{6.2}$$

in axi-symmetric flow. The only flows which strictly satisfies these conditions are the Poiseuille and Couette flows, but boundary layers, jets, wakes and mixing regions are approximately parallel, and have usually been treated as such in stability investigations. The procedure is to consider a small perturbation to the flow cited in (6.1) or (6.2), of the form

$$u = U + u', \qquad v = v', \ w = 0, \tag{6.3}$$

where

$$u' = \frac{\partial \psi}{\partial y},$$

$$v' = -\frac{\partial \psi}{\partial x},$$

$$\psi = \phi(y)\exp(ik(x - ct)), \tag{6.4}$$

real parts being implied, hence restricting all operations on the functions to being linear. It may be shown after the appropriate analysis that $\phi(y)$ satisfies a fourth-order linear differential equation, when square and higher power of disturbance are neglected.

It is worth noting in here that the assumed disturbance, (6.3), is two-dimensional. It is recognized that the flows of the type (6.1) are more unstable to two-dimensional disturbances than to three-dimensional ones. The disturbance (6.4) represents a progressive wave motion, having a wavelength $\lambda = \frac{2\pi}{k}$, where k is called the wavenumber, and the waves are propagated in the x-direction at a speed equal to the real part of c. The crucial matter is whether the imaginary part of c is positive or negative. If positive then the disturbance will grow exponentially as t increases, and if negative the disturbance will decay exponentially as t increases. By solving the aforementioned fourth-order differential equation and examining for which values of the parameters physically acceptable solutions exist, it is possible to predict a Reynolds number for the flow such that at lower Reynolds numbers all small disturbances will be damped, but at higher Reynolds numbers, disturbances exist which can grow spontaneously. This situation is, however, far removed from turbulence. The flow is really a two-dimensional unsteady laminar flow, whereas turbulence is essentially three-dimensional in nature. As a simple example of the kind of situation, we note that the flow between two rotating concentric cylinders is, at a certain relative angular velocity, unstable and changes from the laminar pattern to another laminar pattern with a disturbance of a cellular form. It is only as the relative angular velocity is increased further that turbulent flow is obtained.

6.3 The essential characteristics of turbulence

One of the most essential characteristics of turbulent flow is that the fluctuations, of pressure or velocity say, at a point are not significantly related to fluctuations at a neighbouring point a relatively short distance away. Thus if we consider a point at which a velocity component is instantaneously positive, then clearly the same velocity component at an adjacent point must also be positive, since viscous dissipation limits the magnitudes of the instantaneous velocity gradients which a flow can sustain. At a sufficiently long distance, however, the velocity component can be positive or negative. We may therefore think of a distance, ℓ_c which is such that, there is a good correlation between fluctuations at two points separated by a distance less than ℓ_c, and a poor correlation for points separated by a distance greater than ℓ_c. This distance ℓ_c is called the correlation radius.

We now consider a fluid flow being divided up into a large number of small regions, say spheres of radius ℓ_c. For any two points in the same sphere the respective fluctuations will be well correlated, but for points in different spheres the correlation will be poor. Such volumes of fluid are called correlation volumes or eddies. It is indeed an essential characteristic of turbulence that, in general, there are disturbances having a very wide range of wavelengths. The initial instability of a laminar flow would occur with disturbances of large wavelengths, that is small wave-numbers. Because of the essential nonlinearity of the equations of motions, the various modes which are present will interact, and there will be energy transfer between modes and into new modes, the direction of this energy transfer is from smaller wave-number to higher wave-number, since large-scale disturbances break down into small-scale disturbances. The transfer takes place in what is sometimes called a cascade process, the term rightly implying the successive breakdown into disturbances of smaller and smaller scale. After a sufficiently large number of steps of the cascade, the disturbance of large wave-number will, in effect, be independent of the initial form of the small wave-number disturbances. In other words, there is statistical decoupling between the low wave-number and high wave-number modes. There is a limit of the extent to which this process can continue. As the wave-number increases, so does the viscous dissipation of energy. The range of wave-numbers carrying significant energy is accordingly limited by viscosity.

6.4 Reynolds equations for turbulent motion

Reynolds was the first scientist to put the study of turbulence upon a firm mathematical foundation. To deduce the equations, we begin with the Navier-Stokes equations for incompressible flow, which are valid quite generally for laminar or turbulent flow, namely,

$$\frac{\partial v_i}{\partial x_i} = 0 \tag{6.5}$$

$$\frac{\partial v_i}{\partial t} + \frac{\partial}{\partial x_j}(v_i v_j) = -\frac{1}{\rho}\frac{\partial p}{\partial x_i} + \nu \nabla^2 v_i, \tag{6.6}$$

where rectangular Cartesian coordinates are used, and the suffix notation is used. It is obvious that the convective terms in the momentum equation (6.6) $\frac{\partial}{\partial x_j}(v_i v_j) = v_j \frac{\partial u_i}{\partial x_j} + u_i \frac{\partial v_j}{\partial x_j} = v_j \frac{\partial u_i}{\partial x_j}$. However, since the characteristics of turbulent flow fluctuate rapidly and also randomly, it is useful to regard them as made up of a mean part and a fluctuating part. Hence we define

$$v_i = \overline{v_i} + v_i'$$

$$p = \overline{p} + p', \tag{6.7}$$

where

$$\overline{v_i} = \frac{1}{2T}\int_{-T}^{T} v_i dt$$

$$\overline{p} = \frac{1}{2T}\int_{-T}^{T} p dt$$

Here $\overline{v_i}$ and $\overline{p}$ represent the mean velocities and pressure, respectively. v_i' and p' are the fluctuations about the mean, the averaging time $2T$ being larger than the largest time scale of fluctuations in the turbulence. If the turbulence is not statistically steady, but is decaying with time for example, it will be tacitly assumed that $2T$ is small compared with the time scale of the decay, otherwise the concept of mean velocity or pressure becomes somewhat nebulous. It is worth noting here that the mean of the fluctuating terms are zero, i.e. $\frac{1}{2T}\int_{-T}^{T} v' dt = 0$, $\frac{1}{2T}\int_{-T}^{T} p' dt = 0$.

We now take the means of equations (6.5) and (6.6). There is no difficulty in dealing with the various linear terms, since the process of differentiating and of taking mean values may easily be shown to commute, as follows. For space derivatives we may write

$$\begin{aligned}\overline{\frac{\partial F}{\partial x_i}} &= \frac{1}{2T}\int_{-T}^{T}\frac{\partial F}{\partial x_i}dt \\ &= \frac{\partial}{\partial x_i}\left(\frac{1}{2T}\int_{-T}^{T} F dt\right) \\ &= \frac{\partial \overline{F}}{\partial x_i}\end{aligned}$$

where F is any scalar function. For time derivatives, similarly we write

$$\begin{aligned}\overline{\frac{\partial F}{\partial t}} &= \frac{1}{2T}\int_{-T}^{T}\frac{\partial F}{\partial t}dt \\ &= \frac{\partial}{\partial t}\left(\frac{1}{2T}\int_{-T}^{T} F dt\right) \\ &= \frac{\partial \overline{F}}{\partial t}\end{aligned}$$

Thus, using these formulas the mean values of (6.5) and (6.6) can be written as

$$\frac{\partial \overline{v_i}}{\partial x_i} = 0, \tag{6.8}$$

$$\frac{\partial \overline{v_i}}{\partial t} + \frac{\partial}{\partial x_i}\overline{(v_i v_j)} = -\frac{1}{\rho}\frac{\partial \overline{p}}{\partial x_i} + \nu\nabla^2\overline{v_i}, \tag{6.9}$$

Now let us deal with the nonlinear terms. Using (6.7) it follows that

$$\begin{aligned} v_i v_j &= (\overline{v_i} + v_i')(\overline{v_j} + v_j') \\ &= \overline{v_i}\,\overline{v_j} + \overline{v_i} v_j' + \overline{v_j} v_i' + v_i' v_j' \end{aligned}$$

Upon taking mean values, remembering that all fluctuations have zero mean values, we obtain

$$\overline{v_i v_j} = \overline{v_i}\,\overline{v_j} + \overline{v_i' v_j'}.$$

Hence (6.9) may be written as

$$\begin{aligned} \frac{\partial \overline{v_i}}{\partial t} + \frac{\partial}{\partial x_j}(\overline{v_i}\,\overline{v_j}) &= -\frac{1}{\rho}\frac{\partial \overline{p}}{\partial x_i} + \nu \nabla^2 \overline{v_i} - \frac{\partial}{\partial x_j}\overline{(v_i' v_j')} \\ &= -\frac{1}{\rho}\frac{\partial}{\partial x_j}\left\{ \overline{p}\delta_{ij} - \mu \frac{\partial \overline{v_i}}{\partial x_j} + \rho\overline{(v_i' v_j')} \right\}. \end{aligned} \tag{6.10}$$

Note the striking similarity between the equations (6.5) and (6.6), for the instantaneous pressure and velocities, and equations (6.8) and (6.10), for the mean pressure and velocities, the equations being identical except for the one term. The three terms on the right-hand side of (6.10) represent, respectively, the mean normal pressure, the mean viscous stresses and the mean transfer of momentum by the turbulence. The nine components of the tensor

$$\rho\overline{(v_i' v_j')}$$

are usually referred to as the Reynolds stresses. The fact that these cannot usually be neglected in turbulent flow is alone responsible for the fact that the mean velocity profile in a turbulent flow is very different from that in the associated laminar flow.

In order to see more clearly the significance of the Reynolds stresses, we shall consider the turbulent analogue of the laminar Poiseuille flow between parallel planes in the following section.

6.5 Turbulent flow between parallel planes

In this section we consider Poiseuille's flow in two dimension between two infinite parallel planes. A distance between the planes is $2h$ apart. We take the x_1 axis parallel to the planes, and in the direction of the flow, x_2 normal to the planes and x_3 at right-angles to both x_1 and x_2. We follow the procedure adopted in case of laminar flow discussed before, and we seek solutions in which the mean streamlines are parallel to the planes, so that $\overline{v_2}$ and $\overline{v_3}$ are both zero. Then we consider that $\overline{v_1}$ is a function only of x_2, and t. For statistically steady flow we therefore write

$$\overline{v_1} = \overline{v_1}(x_2), \quad \overline{v_2} = \overline{v_3} = 0. \tag{6.11}$$

It is also assumed that the Reynolds stresses are function of x_2 alone. The continuity equation (6.8) is automatically satisfied by (6.11). Upon substituting into (6.10) from (6.11) many terms disappear identically. Thus, for all values of i, we have

$$\frac{\partial \overline{v_i}}{\partial t} = 0,$$

$$\frac{\partial}{\partial x_j}(\overline{v_i}\,\overline{v_j}) = \frac{\partial}{\partial x_2}(\overline{v_i}\,\overline{v_2}) = 0,$$

$$\frac{\partial}{\partial x_j}\left(\frac{\partial \overline{v_i}}{\partial x_j}\right) = \frac{\partial^2 \overline{v_i}}{\partial x_2^2},$$

and
$$\frac{\partial}{\partial x_j}(\rho(\overline{v_i' v_j'})) = \frac{\partial}{\partial x_2}(\rho\overline{v_i' v_2'}).$$

Since the mean pressure may be assumed not to depend on x_3, it follows that (6.11) yields

$$0 = -\frac{1}{\rho}\frac{\partial \overline{p}}{\partial x_i} + \nu\frac{\partial^2 v_i}{\partial x_2^2} - \frac{\partial}{\partial x_2}(\overline{v_i' v_2'}).$$

Upon setting $i = 1, 2, 3$, respectively, we obtain

$$\frac{\partial \overline{p}}{\partial x_1} + \frac{\partial}{\partial x_2}(\rho\overline{v_1' v_2'}) = \nu\frac{\partial^2 \overline{v_1}}{\partial x_2^2} \tag{6.12}$$

$$\frac{\partial \overline{p}}{\partial x_2} + \frac{\partial}{\partial x_2}(\rho\overline{v_2' v_2'}) = 0 \tag{6.13}$$

$$\frac{\partial}{\partial x_2}(\rho\overline{v_3' v_2'}) = 0 \tag{6.14}$$

Since the Reynolds stresses are independent of x_1 and x_3, we deduce from (6.14) that $\rho\overline{v_3' v_2'}$ is an absolute constant. By virtue of the boundary conditions on either of the planes, $v_2' = v_3' = 0$, it follows that

$$\rho\overline{v_3' v_2'} = 0.$$

Likewise (6.13) shows that $\overline{p} + \rho(\overline{v_2' v_2'})$ is independent of x_2, and since the two terms are separately independent of x_3, it follows that

$$\overline{p} + \rho(\overline{v_2' v_2'}) = F(x_1).$$

The first term of (6.12) is thus a function of x_1 alone, whereas the remaining two terms depend on x_2 alone. Therefore, we have

$$\frac{\partial \overline{p}}{\partial x_1} = -P = \mu\frac{\partial^2 \overline{v_1}}{\partial x_2^2} - \frac{\partial}{\partial x_2}(\rho\overline{v_1' v_2'}), \tag{6.15}$$

Integrating this equation with respect to x_2 partially, we obtain

$$\mu \frac{\partial \overline{v_1}}{\partial x_2} - \rho \overline{v_1' v_2'} = -Px_2 + G(x_1, x_3), \tag{6.16}$$

where $G(x_1, x_3)$ must in fact be a constant, since the other two terms of this equation are dependent on x_2 alone. If we assume that the flow is symmetrical in nature, so that $\partial \overline{v_1}/\partial x_2$ and $\overline{v_1' v_2'}$ are both zero on the plane of symmetry, $x_2 = 0$, it follows that G is zero, and (6.16) may then be written simply

$$\mu \frac{\partial \overline{v_1}}{\partial x_2} = \rho \overline{v_1' v_2'} - Px_2. \tag{6.17}$$

Integrating again with respect to x_2 we have an alternative form

$$\begin{aligned} \overline{v_1}(x_2) &= \overline{v_1}(0) - \frac{P}{2\mu} x_2^2 + \frac{1}{\mu} \int_0^{x_2} \rho \overline{v_1' v_2'} dx_2 \\ &= \frac{P}{2\mu}(h^2 - x_2^2) - \frac{1}{\mu} \int_{x_2}^{h} \rho \overline{v_1' v_2'} dx_2 \end{aligned}$$

In obtaining the last result we used the boundary condition $\overline{v_1}(h) = 0$ such that the unknown constant $\overline{v_1}(0) = \frac{P}{2\mu} h^2 - \frac{1}{\mu} \int_0^h \rho \overline{v_1' v_2'} dx_2$.

Remark

These results, in principle, could be used in either of two ways. First, by making use of experimental measurements, with hot-wire anemometer, of either $\overline{v_1}(x_2)$ or $\overline{v_1' v_2'}$ equation (6.17) could be used to calculate the other. Secondly, if some assumptions or guess be made as to a relationship for $\overline{v_1' v_2'}$ in terms of $\overline{v_1}$, $\frac{\partial \overline{v_1}}{\partial x_2}$ etc., then (6.17) becomes a differential equation for $\overline{v_1}$. Plausible relationships of this nature have been given by Prandtl and other researchers. We pursue this matter in the next section.

6.6 Mixing-length theories of turbulence

Prandtl's momentum-transfer theory

We start with the general ideas behind Prandtl's hypothesis to study the quasi-steady parallel shear flow. We use the (x, y) coordinates with (u, v) as associated velocity components, and consider an element of fluid in a flow with a mean velocity $\overline{u}(y)$ parallel to the x-axis. Due to the general mixing-up which takes place in the turbulent flow, this fluid element is assumed to be carried away a distance ℓ in the transverse direction, conserving its momentum as it does so. The mean velocity of the fluid now surrounding the element is

$$\overline{u}(y + \ell) \approx \overline{u}(y) + \ell \frac{\partial \overline{u}}{\partial y}. \tag{6.18}$$

The element accordingly has an excess momentum, as compared with the surrounding fluid, of amount

$$-\rho\ell\frac{\partial\overline{u}}{\partial y}d\tau,$$

which is assumed to be representative of the turbulent momentum fluctuation

$$\rho u' d\tau.$$

By equating these expressions we deduce that

$$u' \sim -\ell\frac{\partial\overline{u}}{\partial y}. \tag{6.19}$$

Now it is a matter of experimental observation that the turbulent velocity fluctuations have equal orders of magnitude. Accordingly we may deduce from (6.19) that

$$\rho\overline{u'v'} \approx -\rho\ell^2\frac{\partial\overline{u}}{\partial y}\left|\frac{\partial\overline{u}}{\partial y}\right|, \tag{6.20}$$

the sign again being in accord with experimental observation. The length ℓ is called the mixing-length, and is representative of the distance in which an element of fluid mixes completely with the surrounding fluid. The mixing-length is usually determined on the following basis. For flow in a region very close to a rigid wall, since u' and v' are both zero on the wall it is reasonable to assume that

$$\ell \propto \text{distance from the wall.} \tag{6.21}$$

For flow in a jet, wake or mixing region, we take

$$\ell \propto \text{width.} \tag{6.22}$$

For flow in any shear layer, ℓ probably depends on the local mean velocity profile. It has been suggested that the simplest expression, derived from the mean velocity profile and having the dimensions of a length, is

$$\ell \propto \frac{\partial\overline{u}}{\partial y}\Big/\frac{\partial^2\overline{u}}{\partial y^2}. \tag{6.23}$$

It may be noted that (6.21) and (6.22) are, in effect, special cases of (6.23) and are considered with it. The choice of the constant of proportionality in (6.21) to (6.23) is determined empirically by reference to experimental results.

The above hypothesis treats momentum as a transferable property, and is due to Prandtl. There are two other hypotheses which we shall not pursue here. One, which is due to Taylor, treats vorticity as a transferable property, and the other which is due to Prandtl, introduces the concept of a constant turbulent exchange coefficient. Rather we illustrate the general principles behind the use of mixing-length theories by applying the momentum-transfer theory to calculate the flow in a turbulent jet.

The mean velocity in a two-dimensional turbulent jet

We consider the flow of a thin jet into fluid at rest. We shall assume that the boundary layer approximation may be made so that pressure gradients are neglected, and that the viscous stresses are negligible in comparison with the Reynolds stresses. Then in (6.10) we neglect pressure gradients and viscous terms, and by appropriate use of the condition

$$\frac{\partial}{\partial y} >> \frac{\partial}{\partial x},$$

it may be shown, as in deriving the laminar boundary layer equation

$$\overline{u}\frac{\partial \overline{u}}{\partial x} + \overline{v}\frac{\partial \overline{u}}{\partial y} = -\frac{\partial}{\partial y}\overline{(u'v')}, \tag{6.24}$$

and

$$\frac{\partial \overline{u}}{\partial x} + \frac{\partial \overline{v}}{\partial y} = 0. \tag{6.25}$$

In view of the symmetry about the axis of the jet, we consider only the flow for which $y \geq 0$, and since $\frac{\partial \overline{u}}{\partial y} \leq 0$ when $y \geq 0$, it follows from (6.20) that

$$\overline{u'v'} = \ell^2 \left(\frac{\partial \overline{u}}{\partial y}\right)^2, \quad y \geq 0.$$

We introduce a stream function ψ, such that

$$\overline{u} = \frac{\partial \psi}{\partial y} = \psi_y$$

$$\overline{v} = -\frac{\partial \psi}{\partial x} = -\psi_x \tag{6.26}$$

so that (6.25) is automatically satisfied, and (6.24) becomes

$$\psi_y \psi_{yx} - \psi_x \psi_{yy} = -\frac{\partial}{\partial y}\left(\ell^2 \psi_{yy}^2\right). \tag{6.27}$$

We look for a similar solution of the form

$$\psi = x^n f(\eta), \tag{6.28}$$

where $\eta = y/x^m$ so that the width of the jet at any station x is proportional to x^m. The mixing-length ℓ is accordingly taken to be of the form

$$\ell = ax^m,$$

where a is any absolute constant.

By the appropriate differentiation of equation (6.28) it follows that

$$\begin{aligned}
\psi_x &= x^{n-1}(nf - m\eta f'), \\
\psi_y &= x^{n-m} f', \\
\psi_{yy} &= x^{n-2m} f'', \\
\psi_{yyy} &= x^{n-3m} f''', \\
\psi_{yx} &= x^{n-m-1}((n-m)f' - m\eta f''),
\end{aligned} \tag{6.29}$$

where primes here denote derivatives with respect to η, and substitution into (6.27) then yields, after a little reduction,

$$nff'' + (m-n)f'^2 = 2a^2x^{1-m}f'''. \tag{6.30}$$

This equation is self consistent only if

$$m = 1.$$

Further, the condition of constant momentum flux at different station down the jet must be satisfied. This condition follows by integrating (6.24) from $y = -\infty$ to $y = \infty$, yields

$$\frac{d}{dx}\int_{-\infty}^{\infty} \overline{u}^2 dy = 0. \tag{6.31}$$

By virtue of (6.26) and (6.29)

$$\overline{u}^2 = x^{2n-2m}(f')^2,$$

and, by (6.28) since the integral is to be evaluated along a section $x =$ constant, we may write $dy = x^m d\eta$, so that (6.31) yields

$$\frac{d}{dx}\left\{x^{2n-m}\int_{-\infty}^{\infty}(f')^2 d\eta\right\} = 0.$$

Thus, in order that the momentum flux shall be independent of x, we must have $2n - m = 0$, so that $n = \frac{1}{2}$. Upon substituting for m and n the equation (6.30) simply becomes

$$ff'' + f'^2 = 4a^2 f'' f''',$$

the boundary conditions being those of symmetry and of zero mean velocity at the edge of the jet.

This equation may be immediately integrated once to yield

$$ff' = 2a^2(f'')^2,$$

which may be integrated numerically, subject to the conditions of symmetry, $f(0) = f''(0) = f^{iv}(0) = \cdots = 0$, and the zero mean velocity at the edge of the jet, $f' =$

$f'' = f''' = \cdots = 0$. It is found that the conditions at the edge of the jet must be satisfied at

$$\eta = y/x = 3.04a^{2/3}. \tag{6.32}$$

When the value of the constant a has been suitably chosen, comparison with experiment confirms that a turbulent jet spreads linearly in the manner predicted by (6.32), and that the maximum velocity of the jet decreases as $x^{-\frac{1}{2}}$, as the theory also requires.

6.7 Turbulent boundary layers

Two-dimensional turbulent boundary layer

We consider in this section the development of turbulent boundary layer equations in two-dimension with steady mean flow. This may be done by starting from the boundary layer approximation for an arbitrary flow discussed before, namely

$$\frac{\partial u}{\partial x} + \frac{\partial v}{\partial y} = 0,$$

$$\frac{\partial u}{\partial t} + u\frac{\partial u}{\partial x} + v\frac{\partial u}{\partial y} = -\frac{1}{\rho}\frac{\partial p}{\partial x} + \nu\frac{\partial^2 u}{\partial y^2}.$$

This is set of unsteady two-dimensional boundary layer equations. In these equations we write the transformations as

$$u = \overline{u} + u'$$
$$v = \overline{v} + v'$$

and take the mean values. It has been found successively, following the procedure adopted in deriving Reynolds's form of the full motion, that

$$\overline{\frac{\partial u}{\partial t}} = \frac{\partial \overline{u}}{\partial t}$$

which is zero, because the main flow is steady

$$\overline{u\frac{\partial u}{\partial x}} = \overline{u}\frac{\partial \overline{u}}{\partial x} + \frac{1}{2}\frac{\partial}{\partial x}\overline{(u'^2)};$$

$$\begin{aligned}\overline{v\frac{\partial u}{\partial x}} &= \overline{v}\frac{\partial \overline{u}}{\partial y} + \overline{v'\frac{\partial u'}{\partial y}} \\ &= \overline{v}\frac{\partial \overline{u}}{\partial y} + \frac{\partial}{\partial y}\overline{(u'v')} - \overline{u'\frac{\partial v'}{\partial y}} \\ &= \overline{v}\frac{\partial \overline{u}}{\partial y} + \frac{\partial}{\partial y}\overline{(u'v')} + \frac{1}{2}\frac{\partial}{\partial x}\overline{(u'^2)},\end{aligned}$$

upon using the continuity equation. The boundary layer equations for turbulent flow, with a steady mean velocity, then become

$$\begin{aligned}\overline{u}\frac{\partial \overline{u}}{\partial x} + \overline{v}\frac{\partial \overline{u}}{\partial y} &= -\frac{1}{\rho}\frac{\partial \overline{p}}{\partial x} + \nu\frac{\partial^2 \overline{u}}{\partial y^2} - \frac{\partial}{\partial y}\overline{(u'v')} - \frac{\partial}{\partial x}\overline{(u'^2)} \\ &= -\frac{1}{\rho}\frac{\partial \overline{p}}{\partial x} + \frac{1}{\rho}\frac{\partial \tau}{\partial y} - \frac{\partial}{\partial x}\overline{(u'^2)} \end{aligned} \tag{6.33}$$

where $\tau = \mu\frac{\partial \overline{u}}{\partial y} - \rho\overline{u'v'}$. It is usual to neglect the last term on the right-hand side of (6.33), since it is found experimentally that the Reynolds stresses vary only slowly with x, so we have

$$\overline{u}\frac{\partial \overline{u}}{\partial x} + \overline{v}\frac{\partial \overline{u}}{\partial y} = -\frac{1}{\rho}\frac{\partial \overline{p}}{\partial x} + \frac{1}{\rho}\frac{\partial \tau}{\partial y} \tag{6.34}$$

which is identical with the equations for laminar flow, except that τ is the sum of a viscous stress and a Reynolds stress. By examining the form we see that when $y \to \infty$, we define that

$$-\frac{1}{\rho}\frac{\partial \overline{p}}{\partial x} = \overline{U}\frac{d\overline{U}}{dx}.$$

Thus, we have the complete turbulent boundary layer equations together with the continuity equation as

$$\begin{aligned}\frac{\partial \overline{u}}{\partial x} + \frac{\partial \overline{v}}{\partial y} &= 0, \\ \overline{u}\frac{\partial \overline{u}}{\partial x} + \overline{v}\frac{\partial \overline{u}}{\partial y} &= \overline{U}\frac{d\overline{U}}{dx} + \frac{1}{\rho}\frac{\partial \tau}{\partial y}\end{aligned} \tag{6.35}$$

Remark

The boundary layer can be visualized to consist of three regions. (a) If the boundary layer is really turbulent, then the Reynolds stress will be much greater than the viscous stress throughout the greater part of the boundary layer. (b) On the other hand, the Reynolds stress must tend to zero at the wall itself, by virtue of the boundary condition thereon, so there must always be a region very close to the wall in which the viscous stress dominates. This region is usually referred to as the laminar sub-layer. (c) There is, finally, an intermediate region, in which the velocity profile changes gradually from the appropriate to the laminar sub-layer to that appropriate to the fully developed turbulent later.

We now put these ideas on an approximate quantitative basis. In the laminar sub-layer we may write $\tau \approx \mu\frac{\partial u}{\partial y} \approx \tau_w$. Thus after integration we have $u \approx \frac{\tau_w}{\mu}y = \frac{u_\tau^2 y}{\nu}$, or we can write $\frac{u}{u_\tau} = \frac{yu_\tau}{\nu}$, where we have defined $\frac{\tau_w}{\rho} = u_\tau^2$, and u_τ is referred to as the friction velocity.

In the fully turbulent part of the boundary layer we may write

$$\tau \approx -\rho\overline{u'v'}.$$

Now it is a matter of experimental observation that the shearing stress remains reasonably constant in the region close to the wall, particularly when the pressure gradient along the wall is small. Hence using the Prandtl mixing-length ideas, we may write

$$\begin{aligned}\overline{u'v'} &= -\ell^2 \frac{\partial \overline{u}}{\partial y} \left|\frac{\partial \overline{u}}{\partial y}\right| \\ &= -\ell^2 \left(\frac{\partial \overline{u}}{\partial y}\right)^2,\end{aligned}$$

and using relation $\ell = ky$ and also $\frac{\tau_w}{\rho} = u_\tau^2$ we may write

$$\tau_w \approx \rho k^2 y^2 \left(\frac{\partial \overline{u}}{\partial y}\right)^2,$$

close to the wall but in the fully turbulent layer. Upon taking the square root of this equation and rearranging yields

$$\frac{\partial \overline{u}}{\partial y} = \frac{u_\tau}{ky},$$

which integrates to give $\overline{u} = \frac{u_\tau}{k} \ln y +$ constant, and in non-dimensional form it can be written as

$$\frac{\overline{u}}{u_\tau} = A \ln \frac{y u_\tau}{\nu} + B$$

where $A = \frac{1}{k}$.

Figure 11.1 shows a number of experimentally measured mean velocity profiles for turbulent boundary-layer on smooth surface. The results are plotted on a semi-logarithmic scale, and it is clearly seen that when $20 < y u_\tau / \nu < 400$, the mean velocity follows the form given by the above equation, the constant being $A = 2.5$ and $B = 5.5$.

Flow in the absence of pressure gradient

We consider in this section flow in the absence of pressure gradient. The preceding study can be applied to the special case of zero pressure gradient. Let us consider an approach which has proved useful in the study of the turbulent boundary layer on a flat plate. It is a matter of experimental observation that the mean velocity profile of a turbulent boundary layer on a flat plate may be expressed overall by power law

$$\frac{u}{U} = \left(\frac{y}{\delta}\right)^{\frac{1}{n}} = \eta^{\frac{1}{n}} \qquad (6.36)$$

where n is a slowly varying function of the Reynolds number, varying about between 5 and 9 as $R_x = \frac{Ux}{\nu}$ varies between 10^5 and 10^{10}. It is now useful to note that in

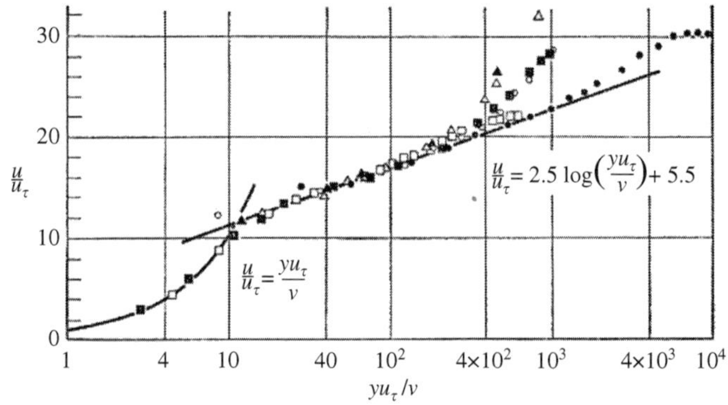

Figure 6.1: Experimental mean velocity profiles. (Adapted from Thwaites, B., *Incompressible Thermodynamics*, Clarendon Press: Oxford, 1960. From [2].)

the remainder of this chapter for convenience we shall omit the bar from the mean velocity $\overline{u}$, which will therefore be denoted simply by u: it is hoped that there should be no ambiguity as a result of this simplified notation. Thus with the profile (6.36) it is easily seen that the displacement thickness of boundary layer in turbulent flow

$$\begin{aligned} \delta_1 &= \int_0^\delta \left(1 - \frac{u}{U}\right) dy \\ &= \delta \int_0^1 \left(1 - \eta^{1/n}\right) d\eta \\ &= \delta \left(1 - \frac{1}{1 + 1/n}\right) = \frac{\delta}{1+n} \end{aligned} \tag{6.37}$$

Likewise we can show that the momentum thickness of turbulent boundary layer is

$$\begin{aligned} \delta_2 &= \int_0^\delta \frac{u}{U}\left(1 - \frac{u}{U}\right) dy \\ &= \frac{n\delta}{(n+1)(n+2)} \end{aligned} \tag{6.38}$$

so that $$H = \frac{\delta_1}{\delta_2} = \frac{n+2}{n}.$$

Thus we may write (6.36) as

$$\frac{u}{U} = \left(\frac{H-1}{H(H+1)}.\frac{y}{\delta_2}\right)^{(H-1)/2}$$

$$\text{or} \qquad \frac{u}{U} = \left(\frac{H-1}{H+1}.\frac{y}{\delta_1}\right)^{(H-1)/2}. \tag{6.39}$$

For flow with zero pressure gradient or small pressure gradient, since the non-dimensional velocity profile (6.36) holds over a good portion of the boundary layer, it must be possible to write (6.36) as $\frac{u}{u_\tau} = C\left(\frac{yu_\tau}{\nu}\right)^{1/n}$, and hence

$$\frac{u}{u_\tau} = C\left(\frac{u_\tau\delta}{\nu}\right)^{1/n} = C^{\frac{n}{n+1}}\left(\frac{U\delta}{\nu}\right)^{\frac{1}{n+1}}. \tag{6.40}$$

Now the momentum integral equation may be written as

$$\frac{d\delta_2}{dx} = \frac{\tau_w}{\rho U^2} = \left(\frac{u_\tau}{U}\right)^2,$$

when there is no pressure gradient. Thus, upon neglecting the small variations of n with x, substitution from (6.38) and (6.40) yields

$$\frac{d\delta}{dx} = \frac{(n+1)(n+2)}{2}C^{-\frac{2n}{n+1}}\left(\frac{U\delta}{\nu}\right)^{-\frac{2}{n+1}},$$

which integrates to give

$$\delta^{\frac{n+3}{n+1}} = \frac{(n+2)(n+1)}{n}C^{-\frac{2n}{n+1}}\left(\frac{U}{\nu}\right)^{-\frac{2}{n+1}}x,$$

Therefore, δ can be evaluated to yield

$$\delta = \left\{\frac{(n+2)(n+1)}{n}\right\}^{\frac{n+1}{n+3}} C^{-\frac{2n}{n+1}} R_x^{-\frac{2n}{n+3}} x$$

$$= DR_x^{-\frac{n}{n+3}} x. \tag{6.41}$$

Having calculated δ, we may use (6.40) to derive u_{tau}, or C_f, where

$$C_f = \frac{\tau_w}{\frac{1}{2}\rho U^2} = 2\left(\frac{u_{tau}}{U}\right)^2.$$

Thus we have

$$
\begin{aligned}
C_f &= 2C^{-\frac{2n}{n+1}}\left(\frac{U\delta}{\nu}\right)^{-\frac{2}{n+1}} \\
&= 2C^{-\frac{2n}{n+1}}\left\{D(R_x)^{\frac{n+1}{n+3}}\right\}^{-\frac{2}{n+1}} \\
&= \frac{2nD}{(n+2)(n+3)}(R_x)^{\frac{2}{n+3}}. \qquad (6.42)
\end{aligned}
$$

Similarly we can calculate

$$
\begin{aligned}
\delta_2 &= \frac{n\delta}{(n+1)(n+)} \\
&= \frac{nD}{(n+1)(n+2)}R_x^{-\frac{2}{n+3}}x. \qquad (6.43)
\end{aligned}
$$

The relationship between C_f and R_x as given by the experimental results is found in Curle and Davies [2] (Table 7.1: Properties of the turbulent boundary layer at constant pressure, p. 251) and will not be repeated here.

6.8 Correlation theory of homogeneous turbulence

In the previous section we have demonstrated various aspects of the mean motion. In this section, we shall turn our attention to a brief consideration of turbulence itself. For simplicity we shall consider turbulence whose mean flow is zero, and which is homogeneous in space. By homogeneous we mean that any mean value, $\overline{F(\mathbf{x})G(\mathbf{x}')}$ of the product of one function F taken at a point $\mathbf{x}$ and a second function G taken at a point $\mathbf{x}'$, will be a function only of the separation $\mathbf{r} = \mathbf{x}' - \mathbf{x}$ of the two points; it will not depend independently upon the two points. As a special case of this definition we deduce that all mean values at a point, such as $\overline{F(\mathbf{x})G(\mathbf{x})}$ will be constant, and hence will have zero derivative. Since the mean flow is zero, the velocity fluctuations are exactly equal to the overall velocity components, and primes need not be used to indicate fluctuations. We assumed throughout that the motion is governed by the Navier-Stokes equations, which we write as

$$\frac{\partial v_i}{\partial x_i} = 0, \qquad (6.44)$$

$$\frac{\partial v_i}{\partial t} + \frac{\partial}{\partial x_j}(v_i v_j) = -\frac{1}{\rho}\frac{\partial p}{\partial x_i} + \nu\nabla^2 v_i. \qquad (6.45)$$

6.8.1 Theoretical development of correlation theory

To develop the correlation theory from the Navier-Stokes equations, we multiply each term in (6.45) by v_i, and then take the mean values. It is implicit that summation over the values $i = 1, 2, 3$ is implied in this process. We obtain the result as follows.

$$\frac{\partial}{\partial t}\left(\frac{1}{2}\overline{v_i^2}\right) + \overline{v_i \frac{\partial}{\partial x_j}(v_i v_j)} = -\frac{1}{\rho}\overline{v_i \frac{\partial p}{\partial x_i}} + \nu\overline{v_i \nabla^2 v_i}. \tag{6.46}$$

In this equation we note the identity

$$2v_i \frac{\partial}{\partial x_j}(v_i v_j) = \frac{\partial}{\partial x_j}(v_i^2 v_j) + v_i^2 \frac{\partial v_j}{\partial x_j}.$$

Since the last term is zero by virtue of continuity equation it follows that

$$\overline{v_i \frac{\partial}{\partial x_j}(v_i v_j)} = \frac{1}{2}\overline{\frac{\partial}{\partial x_j}(v_i^2 v_j)} = 0, \tag{6.47}$$

because of the condition of homogeneity. Likewise we have

$$\overline{v_i \frac{\partial p}{\partial x_i}} = \frac{\partial}{\partial x_i}\overline{(p v_i)} - \overline{p \frac{\partial v_i}{\partial x_i}} = 0, \tag{6.48}$$

the two terms being zero by virtue of homogeneity and continuity, respectively. Again, we may write

$$\begin{aligned}
\overline{v_i \nabla^2 v_i} &= \overline{v_i \frac{\partial}{\partial x_j}\left(\frac{\partial v_i}{\partial x_j}\right)} \\
&= \overline{\frac{\partial}{\partial x_j}\left(v_i \frac{\partial v_i}{\partial x_j}\right)} - \overline{\left(\frac{\partial v_i}{\partial x_j}\right)^2} \\
&= -\overline{\left(\frac{\partial v_i}{\partial x_j}\right)^2}.
\end{aligned} \tag{6.49}$$

This is also by virtue of the homogeneity condition. Upon substituting from (6.47) to (6.49), we see that (6.46) becomes

$$\frac{\partial}{\partial t}\left(\frac{1}{2}\overline{v_i^2}\right) = -\nu\overline{\left(\frac{\partial v_i}{\partial x_j}\right)^2}. \tag{6.50}$$

This equation has a simple physical interpretation. It is that the rate of change of mean kinetic energy is determined by the mean rate of dissipation. We may alternatively interpret this equation in terms of mean-square vorticity. We also note that the three components of vorticity are of the form

$$\frac{\partial v_i}{\partial x_j} - \frac{\partial v_j}{\partial x_i}$$

for appropriately chosen i and j. Thus if we consider

$$\left(\frac{\partial v_i}{\partial x_j} - \frac{\partial v_j}{\partial x_i}\right)^2,$$

there are six non-zero terms, the three terms where $i = j$ being each zero, the non-zero terms being equal, respectively, to the squares of plus and minus each of the three components of vorticity. Thus the mean square vorticity is

$$\omega^2 = \omega_i^2 = \frac{1}{2}\left(\frac{\partial v_i}{\partial x_j} - \frac{\partial v_j}{\partial x_i}\right)^2.$$

We expand this equation and it is interesting to see that the two squared terms are identically equal and hence we have

$$\begin{aligned}\omega^2 &= \left(\frac{\partial v_i}{\partial x_j}\right)^2 - \frac{\partial v_i}{\partial x_j}\cdot\frac{\partial v_j}{\partial x_i}\\ &= \left(\frac{\partial v_i}{\partial x_j}\right)^2 - \frac{\partial}{\partial x_j}\left(v_i\frac{\partial v_j}{\partial x_i}\right),\end{aligned}$$

by using the continuity equation. By taking mean values, we find that

$$\overline{\omega^2} = \overline{\left(\frac{\partial v_i}{\partial x_j}\right)^2},$$

in the case of homogeneous turbulence. Hence we may write (6.50) as

$$\frac{\partial}{\partial t}\left(\frac{1}{2}\overline{v_i^2}\right) = -\nu\overline{\omega_i^2} = -\nu\overline{\omega^2}. \tag{6.51}$$

In the following we will obtain a somewhat more general formula than (6.50). In doing so, we introduce a velocity covariance tensor, R_{ij}, defined as follows.

$$\begin{aligned}\overline{v_i(\mathbf{x})v_j(\mathbf{x}')} &= R_{ij}(\mathbf{x}' - \mathbf{x}),\\ \text{or}\qquad \overline{v_i v_j} &= R_{ij}(\mathbf{r}),\end{aligned} \tag{6.52}$$

where the notation is almost self-explanatory. It is fairly clear from this definition that

$$R_{ij}(\mathbf{r}) = R_{ji}(\mathbf{r}) = R_{ij}(-\mathbf{r}) = R_{ji}(-\mathbf{r}).$$

Also by using the continuity equation, we have

$$\frac{\partial}{\partial x'_j}(v_i v'_j) = v_i\frac{\partial v'_j}{\partial x'_j} = 0,$$

remembering that unprimed quantities are not functions of the primed coordinate. Thus we have

$$\frac{\partial}{\partial x'_j}\overline{(v_i v'_j)} = \frac{\partial}{\partial r_j}(R_{ij}) = 0. \tag{6.53}$$

In order to obtain an equation for R_{ij} we write down the momentum equation (6.45) twice, once for v_i and once for v_j as follows:

$$\frac{\partial v_i}{\partial t} + \frac{\partial}{\partial x_k}(v_i v_k) = -\frac{1}{\rho}\frac{\partial p}{\partial x_i} + \nu\nabla^2 v_i,$$

and

$$\frac{\partial v'_j}{\partial t} + \frac{\partial}{\partial x'_k}(v'_j v'_k) = -\frac{1}{\rho}\frac{\partial p'}{\partial x'_j} + \nu\nabla^2 v'_j.$$

We multiply the first of these equations by v'_j, the second by v_i and add, and take mean values. This yields

$$\frac{\partial}{\partial t}\overline{(v_i v'_j)} + \frac{\partial}{\partial x_k}\overline{(v_i v_k v'_j)} + \frac{\partial}{\partial x'_k}\overline{(v_i v'_j v'_k)} + \frac{1}{\rho}\left(\overline{v'_j\frac{\partial p}{\partial x_i}} + \overline{v_i\frac{\partial p'}{\partial x'_j}}\right)$$
$$= \nu\left(\overline{v_i\nabla'^2 v'_j} + \overline{v'_j\nabla^2 v_i}\right). \tag{6.54}$$

We have noticed in (6.54) that

$$\nabla^2(v_i v'_j) = v'_j\nabla^2 v_i = v_i\nabla^2 v'_j.$$

Therefore we may write

$$\frac{\partial}{\partial t}(R_{ij}) = T_{ij} + P_{ij} + 2\nu\nabla^2 R_{ij}, \tag{6.55}$$

where

$$T_{ij} = -\frac{\partial}{\partial x_k}\overline{(v_i v_k v'_j)} - \frac{\partial}{\partial x'_k}\overline{(v_i v'_j v'_k)}$$
$$= \frac{\partial}{\partial r_k}\left(\overline{v_i v_k v'_j} - \overline{v_i v'_j v'_k}\right) \tag{6.56}$$

and

$$P_{ij} = -\frac{1}{\rho}\left(\overline{v'_j\frac{\partial p}{\partial x_i}} + \overline{v_i\frac{\partial p'}{\partial x'_j}}\right)$$
$$= \frac{1}{\rho}\left(\frac{\partial}{\partial r_i}\overline{(p v'_j)} - \frac{\partial}{\partial r_j}\overline{(p' v_i)}\right). \tag{6.57}$$

In due course we shall show that when the turbulence is isotropic, P_{ij} is identically zero. For this reason it was once believed that the term P_{ij} represented a tendency towards isotropy, although it is not now believed that any built-in tendency to isotropy, exists. The term T_{ij} represents energy transfer between different modes of the spectrum. Finally the term $2\nu\nabla^2 R_{ij}$ represents dissipation of energy due to viscosity.

We now turn our attention to the spectral case in which the turbulence, already assumed to be homogeneous, is also isotropic, i.e. it has no directional preferences. This at once simplifies the theory considerably; at the same time, approximately isotropic turbulence can easily be produced on a laboratory scale, for example when a uniform stream of fluid passes through a regular grid in a wind-tunnel, so any predictions of the theory can be tested experimentally.

6.8.2 Isotropic turbulence

In this section we shall investigate the special forms which R_{ij}, T_{ij}, P_{ij}, take when the turbulence is isotropic. To find these forms, we consider an arbitrary vector $Q_i(\mathbf{r})$ and an arbitrary reference vector a_i. Then if Q_i is isotropic, the scalar product $\mathbf{Q} \cdot \mathbf{a} = Q_i a_i$ will depend on a^2, r^2 and $\mathbf{a} \cdot \mathbf{r}$ but not upon $\mathbf{a}$ or $\mathbf{r}$ independently. But $Q_i a_i$ is linear in a_i, so it follow that

$$Q_i a_i = A(r) a_i r_i,$$

$$\text{or} \qquad Q_i = A(r) r_i,$$

since this result must hold for all a_i. If, in addition,

$$\frac{\partial}{\partial r_i}(Q_i) = 0,$$

so that the isotropic vector Q_i is solenoidal, then

$$A(r)\frac{\partial r_i}{\partial r_i} + A'(r)\frac{\partial r}{\partial r_i} r_i = 0,$$

or, since the repeated suffix implies summation,

$$3A + rA' = 0.$$

Thus, upon integration, we have

$$A(r) = C/r^3,$$

where C is a constant. Now $\overline{pv_i}$ satisfies all these conditions. It is an isotropic vector function of r, and since $(\frac{\partial}{\partial x_i'})(\overline{pv_i'})$ is zero by the continuity equation, it follows that

$$\frac{\partial}{\partial r_i}(\overline{pv_i'}) = 0,$$

that is $\overline{pv_i'}$ is solenoidal. Thus $\overline{pv_i'}$ is of the form Cr^{-3}, and if this is to be physically reasonable as $r \to 0$, we must have $C = 0$, and hence $\overline{pv_i'} = 0$ in isotropic turbulence. It follows from (6.57) that

$$P_{ij} = 0. \tag{6.58}$$

In a similar manner we consider an isotropic tensor function $Q_{ij}(\mathbf{r})$. The function $Q_{ij}a_ib_j$ must be dependent only on a^2, b^2, r^2, $\mathbf{a}\cdot\mathbf{r}$, $\mathbf{b}\cdot\mathbf{r}$ and $\mathbf{a}\cdot\mathbf{b}$ but not $\mathbf{a}$, $\mathbf{b}$, $\mathbf{r}$ independently. However, it is linear in both a and b, so

$$\begin{aligned} Q_{ij}a_ib_j &= A(r)(a_ir_i)(b_jr_j) + B(r)a_ib_i \\ &= A(r)(a_ir_i)(b_jr_j) + B(r)a_ib_j\delta_{ij} \end{aligned}$$

where A, B are functions of r^2, that is even function of r. This result holds for all $\mathbf{a}$ and $\mathbf{b}$, so it follows that

$$Q_{ij} = A(r)r_ir_j + B(r)\delta_{ij}, \tag{6.59}$$

if, in addition,

$$\frac{\partial}{\partial r_j}(Q_{ij}) = 0$$

for all i, so that Q_{ij} is solenoidal, then (6.59) yields

$$\begin{aligned} 0 &= A'r_ir_j\frac{\partial r}{\partial r_j} + A\delta_{ij}r_j + Ar_i\frac{\partial r_j}{\partial r_j} + B'\frac{\partial r}{\partial r_j}\delta_{ij} \\ &= A'rr_i + Ar_i + 3Ar_i + B'r_i/r \end{aligned}$$

so that

$$A'r^2 + 4rA + B' = 0. \tag{6.60}$$

Now R_{ij} satisfies all these conditions, so it may be expressed in the form (6.59) with A and B related by (6.60). We introduce now the longitudinal velocity correlation coefficient, $f(r)$ defined by

$$R_{11}(r, 0, 0) = u^2f(r), \tag{6.61}$$

where

$$u^2 = \overline{v_1^2} = \overline{v_2^2} = \overline{v_3^2}.$$

This involves a correlation between the velocity components at two points, each component being parallel to the vector separation of the points. It follows, by (6.59), that

$$R_{11}(r, 0, 0) = r^2A + B, \tag{6.62}$$

and we may solve (6.60) to (6.62) to yield A and B in terms of $f(r)$.

Example 6.1

Deduce that

$$R_{ij}(\mathbf{r}) = u^2\left\{\left(f + \frac{1}{2}rf'\right)\delta_{ij} - \frac{1}{2r}f'r_ir_j\right\}. \tag{6.63}$$

Solution

Given that $R_{11}(r, 0, 0) = u^2 f(r) = A(r)r^2 + B(r)$. Also we have $A'r^2 + 4rA + B' = 0$. To obtain the values of A and B in terms of $f(r)$, we differentiate the equation

$$u^2 f(r) = A(r)r^2 + B(r)$$

with respect to r yielding

$$r^2 A' + 2rA + B' = u^2 f'(r).$$

The values of A and B are given by

$$A(r)r^2 = -u^2 \left[\frac{1}{2} r f'(r)\right]$$

$$B(r) = u^2 \left[\left(f + \frac{1}{2} r f'(r)\right)\right]$$

Hence, we have

$$R_{11}(r, 0, 0) = u^2 \left\{\left(f + \frac{1}{2} r f'\right) - \frac{1}{2} r f'\right\}.$$

Now to determine the general solution for $R_{ij}(\mathbf{r})$), we modify the values of A and B as follows.

$$A(r) r_i r_j = -u^2 \left[\frac{1}{2r} f' r_i r_j\right]$$

$$B(r)\delta_{ij} = u^2 \left[\left(f + \frac{1}{2} r f'\right)\right] \delta_{ij}$$

Hence the solution for $R_{ij}(\mathbf{r})$ is given by

$$R_{ij}(\mathbf{r}) = u^2 \left\{\left(f + \frac{1}{2} r f'\right) \delta_{ij} - \frac{1}{2r} r_i r_j f'\right\}$$

which is the required solution. Note that $r^2 = r_i r_i = r_j r_j$. But $r_i r_j$ is a tensor and has nine terms.

Let us introduce the lateral velocity correlation coefficient, $g(r)$, defined by

$$R_{11}(0, r, 0) = u^2 g(r),$$

which involves the velocity components at two points, each component being normal to the vector separation of the points. We deduce from (6.63) that

$g(r) = f + \frac{1}{2}rf'$, so that

$$\begin{aligned}\int_0^\infty r^n g dr &= \int_0^\infty r^n \left(f + \frac{1}{2}rf'\right) dr \\ &= \int_0^\infty \left\{\frac{d}{dr}\left(\frac{1}{2}r^{n+1}f\right) + \frac{1-n}{2}r^n f\right\} dr \\ &= \left[\frac{1}{2}r^{n+1}f\right]_0^\infty + \frac{1}{2}(1-n)\int_0^\infty r^n f dr \\ &= \frac{1}{2}(1-n)\int_0^\infty r^n f dr, \end{aligned} \tag{6.64}$$

provided $f(r) \to 0$ sufficiently rapidly as $r \to \infty$. It is found experimentally that $f(r)$ is positive for all r, so the integral on the right-hand side of (6.64) is positive, and

$$\begin{aligned}\int_0^\infty r^n g(r) dr &> 0 \text{ if } n < 1 \\ &= 0 \text{ if } n = 1 \\ &< 0 \text{ if } n > 1.\end{aligned}$$

To satisfy these conditions, $g(r)$ must have a negative loop for large r, and the general forms of $f(r)$ and $g(r)$ must be as shown in Fig. 11.2.

Eddy sizes and energy dissipation

In this section we now introduce two scales of turbulence. The first is called the longitudinal integral scale, and is defined by

$$\ell = \int_0^\infty f(r) dr.$$

We shall later demonstrate that this scale is representative of the energy-bearing eddies. To define the second scale we write

$$f(r) = f(0) + \frac{1}{2}f''(0)r^2 + \cdots,$$

for small r. Note here that $f'(0) = 0$. This expansion is appropriate, since $f(r)$ is a even function. Since $2\overline{vv'} \leq \overline{v^2} + \overline{v'^2}$, with equality only occurring when $v \equiv v'$, we note that $f''(0) < 0$, so we write

$$f''(0) = -\frac{1}{\lambda^2}$$

whence

$$f(r) = 1 - \frac{r^2}{2\lambda^2} + \cdots$$

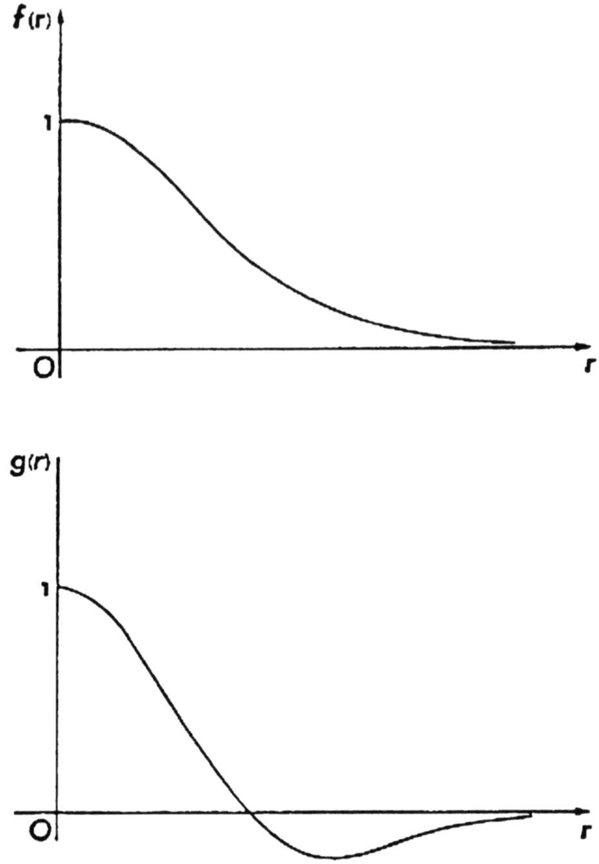

Figure 6.2: Basic velocity correlation functions $f(r)$ and $g(r)$ (from [2]).

It may likewise be shown that when r is small the function $g(r)$ may be expanded as

$$g(r) = 1 - \frac{r^2}{\lambda^2} + \cdots .$$

The rate of energy dissipation, appearing in (6.50), may conveniently be expressed in terms of the length λ, as follows. We note that

$$\begin{aligned}\overline{\left(\frac{\partial v_1}{\partial x_1}\right)^2} &= \lim_{r\to 0} \overline{\left(\frac{\partial v_1}{\partial x_1} \cdot \frac{\partial v_1'}{\partial x_1'}\right)} \\ &= -\lim_{r_1\to 0} \left\{\frac{\partial^2}{\partial r_1^2}\overline{(v_1 v_1')}\right\}\end{aligned}$$

$$= -u^2 f''(0)$$
$$= \frac{u^2}{\lambda^2}.$$

Similarly

$$\overline{\left(\frac{\partial v_1}{\partial x_2}\right)^2} = -u^2 g''(0)$$
$$= \frac{2u^2}{\lambda^2}.$$

So the dissipation rate per unit mass is

$$\begin{aligned}\epsilon &= \nu \overline{\left(\frac{\partial v_i}{\partial x_j}\right)^2} \\ &= \nu \left\{3 \text{ terms like} \overline{\left(\frac{\partial v_1}{\partial x_1}\right)^2} + 6 \text{ terms like} \overline{\left(\frac{\partial v_1}{\partial x_2}\right)^2}\right\} \\ &= \frac{15\nu u^2}{\lambda^2}.\end{aligned}$$

Accordingly equations (6.50) and (6.51) may be written

$$\frac{d}{dt}(u^2) = -\frac{10\nu u^2}{\lambda^2} \tag{6.65}$$

and

$$\overline{\omega^2} = \frac{15u^2}{\lambda^2}. \tag{6.66}$$

Let us now express in approximate quantitative form the idea that the dissipation rate is equal to the rate of break-up of the energy-bearing eddies. Since ℓ is a characteristic length of these eddies, we have

$$-\frac{d}{dt}\left(\frac{3}{2}u^2\right) \approx \frac{3}{2}u^2/(\ell/u),$$

and hence

$$-\frac{d}{dt}(u^2) = \frac{10\nu u^2}{\lambda^2} \approx \frac{u^3}{\ell}. \tag{6.67}$$

Experiments indicate that the right-hand side of (6.67) should be multiplied by a factor which varies only slightly with experimental conditions within the range 0.8 to 1.3.

For the sort of turbulence that is produced under laboratory conditions, when a stream of velocity U impinges on grids of mesh M, it is found that

$$R_\ell = \frac{u\ell}{\nu} \approx 0.01\frac{UM}{\nu} = 0.01R_M.$$

Hence, since (6.67) shows that

$$\frac{\ell^2}{\lambda^2} \approx 0.1 R_\ell$$

it follows that

$$\frac{\ell^2}{\lambda^2} \approx 0.001 R_\ell.$$

Since typical R_M lies in the range from 2,000 to 100,000, it follows that $(\ell/\lambda)^2$ will lie in the range 2–100, and so (ℓ/λ) will be between 1 and 10.

The momentum equation in isotropic turbulence

It is clear from (6.58) that $P_{ij} = 0$ in isotropic turbulence, and hence the momentum equation (6.55) can be written as

$$\frac{\partial}{\partial t}(R_{ij}) - 2\nu\nabla^2 R_{ij} = T_{ij}, \tag{6.68}$$

where, by (6.56) we may write

$$T_{ij} = \frac{\partial}{\partial r_k}\left(S_{ikj}(\mathbf{r}) - S_{jki}(-\mathbf{r})\right) \tag{6.69}$$

with

$$S_{ikj} = \overline{v_i v_k v'_j} \tag{6.70}$$

Since $\frac{\partial}{\partial r_k}\left(S_{ikj}(\mathbf{r})\right)$ is an isotropic function of $\mathbf{r}$, it is of the form $r_i r_j A(r) + \delta_{ij} B(r)$. Thus it is unchanged if i and j are interchanged or when the sign of $\mathbf{r}$ is changed. It follows that (6.69) becomes

$$T_{ij} = 2\frac{\partial}{\partial r_k}\left(S_{ikj}(\mathbf{r})\right).$$

Now the three terms appearing in (6.68) are all solenoidal. The left-hand side has previously been shown to be solenoidal, and the right-hand side is solenoidal since

$$\frac{\partial}{\partial x'_j}\overline{(v_i v_k v'_j)} = 0.$$

Thus the three terms may be expressed in the forms

$$R_{ij} = u^2\left\{\left(f + \frac{1}{2}rf'\right)\delta_{ij} - \frac{1}{2r}f' r_i r_j\right\}, \tag{6.71}$$

$$\nabla^2 R_{ij} = u^2\left\{\left(s + \frac{1}{2}rs'\right)\delta_{ij} - \frac{1}{2r}s' r_i r_j\right\}, \tag{6.72}$$

$$\frac{\partial}{\partial r_k}(S_{ikj}) = u^2\left\{\left(t + \frac{1}{2}rt'\right)\delta_{ij} - \frac{1}{2r}t' r_i r_j\right\}, \tag{6.73}$$

where f, s, t are even functions of r. We deduce s in terms of f by using (6.71) and (6.72) and taking special cases. For example, by setting $i = j$, we have

$$\begin{aligned} u^2(3s + rs') &= \nabla^2 R_{ii} \\ &= \frac{u^2}{r^2}\frac{\partial}{\partial r}\left\{r^2\frac{\partial}{\partial r}(3f + rf')\right\}. \end{aligned}$$

This leads to the result

$$s = f'' + \frac{4}{r}f'. \tag{6.74}$$

Note that $\nabla^2 = \frac{1}{r^2}\frac{\partial}{\partial r}\left(r^2\frac{\partial}{\partial r}\right)$ in spherical polar coordinates.

We next examine equation (6.73), and seek to learn something of the form of t, which clearly may be related to a basic triple velocity correlation. We note first that S_{ikj} is of the form

$$S_{ikj} = r_i r_j r_k A(r) + r_i S_{kj} B(r) + r_k S_{ij} C(r) + r_j \delta_{ik} D(r). \tag{6.75}$$

As usual, these four functions, A, B, C, D, can be reduced to one. For instance, we note from (6.70) that S_{ikj} is symmetrical in i and k, so that

$$B(r) = C(r).$$

Also, since S_{ikj} is solenoidal, we have

$$\frac{\partial}{\partial r_j}(S_{ikj}) = 0,$$

and after performing the relevant algebra this yields

$$0 = r_i r_k\left(rA' + 5A + \frac{2B'}{r}\right) + \delta_{ik}(2B + rD' + 3D).$$

Since this must hold for all values of i and k, we have

$$\begin{aligned} r^2A' + 5rA + 2B' &= 0 \\ \text{and} \qquad 2B + rD' + 3D &= 0. \end{aligned}$$

Upon solving for A and B in terms of D, we obtain

$$\begin{aligned} A &= D'/r \\ B = C &= -\frac{1}{2}(rD' + 3D). \end{aligned} \tag{6.76}$$

As an alternative, we may relate A, B, C, D, to one of the basic two point triple velocity correlations. For example, suppose we write

$$u^3h(r) = S_{112} = \overline{v_1^2(0,0,0)v_2(0,r,0)},$$

then (6.75) shows that

$$u^3 h(r) = rD(r), \tag{6.77}$$

indicating that $h(r)$ is an odd function . We substitute in (6.75) from (6.76), and obtain the result

$$S_{ikj} = u^3 \left\{ \frac{rh' - h}{r^3} r_i r_j r_k - \frac{rh' + 2h}{2r}(r_k \delta_{ij} + r_i \delta_{kj}) + \frac{h}{r} r_j \delta_{ik} \right\} \tag{6.78}$$

Example 6.2

Deduce from (6.73) and (6.78), by considering a special case, that

$$t = -u(h' + 4h/r) \tag{6.79}$$

Solution

This problem can be solved by using the following hint. We do not want to pursue its detailed calculations. It is left to the reader as an exercise. [Hint: Examine the value of $\frac{\partial}{\partial r_k}(S_{iki})$.]

We have now obtained expressions for the scalar functions s and t corresponding to the terms in (6.68), which may therefore be written as

$$\frac{\partial}{\partial t}(u^2 f) - 2\nu u^2 s = 2u^2 t.$$

When we substitute for s from (6.74) and t from (6.79), this becomes

$$\frac{\partial}{\partial t}(u^2 f) - 2\nu u^2 \left(f'' + \frac{4}{r} f' \right) = -2u^3 \left(h' + \frac{4}{r} h \right). \tag{6.80}$$

This equation was first derived by Kármán and Howarth.

Some deductions from the Kármán and Howarth

This section is devoted in deriving certain forms of the Kármán - Howarth equation, to illustrate various forms of isotropic turbulence.

We begin by letting $r \to 0$ in t(6.80). Since $f(r)$ is an even function of r, and $h(r)$ is an odd function of r, it follows that

$$\frac{4}{r} f'(r) \to 4f''(0)$$

$$\text{and} \qquad \frac{4}{r} h(r) \to 4h'(0).$$

Thus, in the limits, (6.80) yields

$$\frac{\partial}{\partial t}(u^2) + 10\nu \left(\frac{u^2}{\lambda^2} \right) = -10u^2 h'(0).$$

But, in (6.65), the left-hand side is exactly zero. Hence we have

$$h'(0) = 0. \tag{6.81}$$

We now differentiate (6.80) twice with respect to r and then put $r = 0$. Proceeding as before we obtain the result

$$\frac{\partial}{\partial t}\left(-\frac{u^2}{\lambda^2}\right) - \frac{14}{3}\nu u^2 f^{iv}(0) = -\frac{14}{3}u^3 h'''(0).$$

This equation may be integrated in terms of mean-square velocity. Using (6.66) we see that it may be written as

$$\frac{d}{dt}\overline{(\omega^2)} = 70u^3 h'''(0) - 70\nu u^2 f^{iv}(0).$$

The two terms on the right-hand side of this equation show how the mean-square vorticity changes. The first term which is positive, represents the fact that turbulence tends to stretch the vortex lines, causing a greater vorticity in a smaller volume and hence a greater mean-square vorticity. The second term represents the dissipation effects of viscosity, must of course be negative. We deduce from the signs of these terms that

$$f^{iv}(0) > 0, \text{ and} \qquad h'''(0) > 0.$$

Since h is an odd function of r, $h(0)$ and $h''(0)$ are both zero, and we have already shown that $h'(0)$ is zero. We therefore deduce that $h'(r) > 0$ for small values of r. Experimentally it is found that this is true, and indeed that $h(r)$ is positive for all values of r.

We now multiply equation (6.80) by r^4 and integrate from 0 to ∞. Thus we obtain

$$\frac{\partial}{\partial t}\left\{\int_0^\infty u^2 f(r) r^4 dr\right\} = 2\nu u^2 \int_0^\infty \frac{\partial}{\partial r}\left(r^4\frac{\partial f}{\partial r}\right) dr - 2u^3 \int_0^\infty \frac{\partial}{\partial r}(r^4 h) dr,$$

or
$$\frac{\partial}{\partial t}\left\{u^2\int_0^\infty r^4 f dr\right\} = 2\nu u^2\left[r^4\frac{\partial f}{\partial r}\right]_0^\infty - 2u^3[r^4 h]_0^\infty. \tag{6.82}$$

Now the first term on the right-hand side is known experimentally to be zero. If the second term were also zero, then the expression $u^2\int_0^\infty r^4 f dr$ would be invariant with time. This expression id referred to as Loitsianskii's invariant. Because of the factor r^4 in the integral the suggestion arose of some measure of permanence in those eddies which contribute to $f(r)$ for large varus of r, namely the large eddies, but this interesting possibility is invalidated because the last term in (6.82) is no longer believed to vanish at infinity.

Finally, we integrate (6.80) itself. This yields

$$\frac{\partial}{\partial t}\left\{u^2\int_0^\infty f(r)dr\right\} = 2\nu u^2\int_0^\infty\left(f'' + \frac{4}{r}f'\right)dr - 2u^2\int_0^\infty\left(h' + \frac{4h}{r}\right)dr$$

or
$$\frac{\partial}{\partial t}(u^2\ell) = 8\nu u^2\int_0^\infty \frac{1}{r}f'(r)dr - 8u^3\int_0^\infty \frac{1}{r}h(r)dr. \tag{6.83}$$

By making suitable assumptions regarding this equation, it may be solved simultaneously with the empirical relationship (6.67). We note that the left-hand side of (6.83) is of order u^3, as (6.67) shows, whereas the first term on the right-hand side is of the order $8u^3(R_\lambda)^{-1}$; accordingly we may neglect this latter term, provided R_λ is large enough. If, further, we assume that the second integral in (6.83) varies only slowly with time, and therefore write

$$\int_0^\infty \frac{1}{r}h(r)dr = C,$$

then the equation approximates to

$$\frac{\partial}{\partial t}(u^2\ell) = -Cu^3.$$

Upon solving simultaneously with (6.67) we obtain the result

$$\begin{aligned} u &= u_0(t-t_0)^{-n} \\ \ell &= \frac{ku_0}{2n}(t-t_0)^{1-n}, \end{aligned} \tag{6.84}$$

where u_0 and t_0 are arbitrary and

$$n = (3 - 2C/k)^{-1}.$$

Example 6.3

Given that the vector $Q_i = A(r)r_i$ is solenoidal, that is Q_i satisfies the continuity equation. Deduce the value of the function $A(r)$ in the form $A(r) = C/r^3$, where C is an arbitrary constant.

Solution

Since Q_i is solenoidal, we have

$$\frac{\partial Q_i}{\partial r_i} = 0.$$

This implies that

$$\frac{\partial}{\partial r_i}(A(r)r_i) = 0$$

$$\text{or} \qquad A'(r)\left(\frac{\partial r}{\partial r_i}\right)r_i + A(r)(\frac{\partial r_i}{\partial r_i}) = 0$$

$$\text{or} \qquad A'(r)\left(\frac{r_i}{r}\right)r_i + A(r)\delta_{ii} = 0$$

The last equation can be written as a first-order ordinary differential equation

$$r\frac{dA}{dr} + 3A = 0$$

the solution of which is simply $A(r) = C/r^3$, where C is an arbitrary constant. Therefore $Q_i = (C/r^3)r_i$. Hence the required result.

Example 6.4

If $Q_{ij} = A(r)r_ir_j + B(r)\delta_{ij}$ is solenoidal, then prove that A and B satisfy this ordinary differential equation $A'r^2 + 4rA + B' = 0$.

Proof

Given that $Q_{ij}(\mathbf{r})$ is solenoidal with respect to i and j. Therefore, we must have by the continuity equation $\frac{\partial Q_{ij}}{\partial r_j} = 0$. Thus we obtain

$$\frac{\partial}{\partial r_j}\left(A(r)r_ir_j + B(r)\delta_{ij}\right) = 0$$

or

$$A'(r)\left(\frac{\partial r}{\partial r_j}\right)r_ir_j + A(r)\left(\frac{\partial r_i}{\partial r_j}\right)r_j + A(r)r_i\left(\frac{\partial r_j}{\partial r_j}\right) + B'(r)\left(\frac{\partial r}{\partial r_j}\right)\delta_{ij} = 0.$$

It is noted that $\frac{\partial r}{\partial r_j} = \frac{r_j}{r}$, $\frac{\partial r_i}{\partial r_j} = \delta_{ij}$ and $\frac{\partial r_j}{\partial r_j} = \delta_{jj}$. Hence using these entities, we obtain

$$A'\left(\frac{r_j}{r}\right)r_ir_j + A\delta_{ij}r_j + Ar_i\delta_{jj} + B'\left(\frac{r_j}{r}\right)\delta_{ij} = 0,$$

which is

$$rA'r_i + A\delta_{ii}r_i + 3Ar_i + \left(\frac{B'}{r}\right)r_i = 0.$$

Using the tensor identity $A\delta_{jj} = 3A$ and then equating the coefficient of r_i to zero, the last equation can be written as

$$rA' + 4A + B'/r = 0.$$

Example 6.5

Given that $S_{ikj}(\mathbf{r}) = r_ir_jr_kA(r) + r_i\delta_{kj}B(r) + r_k\delta_{ij}C(r) + r_j\delta_{ik}D(r)$ is solenoidal, then show that A, B, C, D are related to the ordinary differential equations

$$r^2A' + 5rA + 2B' = 0$$

$$2B + rD'3D = 0.$$

Solution

Given that $S_{ikj}(\mathbf{r}) = \overline{v_i v_k v'_j}$ is a symmetric tensor in i and k. Therefore, $r_i \delta_{kj} B(r) = r_k \delta_{ij} = B(r) = r_k \delta_{ij} C(r)$, and hence we have $B(r) = C(r)$. Thus the given tensor can be rewritten as

$$S_{ikj} = r_i r_j r_k A(r) + (r_i \delta_{kj} + r_k \delta_{ij}) B(r) + r_j \delta_{ik} D(r).$$

This tensor is solenoidal, which means that it satisfies the continuity equation. Thus $\frac{\partial}{\partial r_j}(S_{ikj}) = 0$. Taking the partial derivative with respect to r_j we obtain

$$\begin{aligned}
&\left(\frac{\partial r_i}{\partial r_j}\right) r_j r_k A + r_i \left(\frac{\partial r_j}{\partial r_j}\right) r_k A + r_i r_j \left(\frac{\partial r_k}{\partial r_j}\right) A + r_i r_j r_k A' \left(\frac{\partial r}{\partial r_j}\right) \\
&+ \left\{\left(\frac{\partial r_i}{\partial r_j}\right) \delta_{kj} + \left(\frac{\partial r_k}{\partial r_j}\right) \delta_{ij}\right\} B + \left(r_i \delta_{kj} + r_k \delta_{ij}\right) B' \left(\frac{\partial r}{\partial r_j}\right) \\
&\times \left(\frac{\partial r_j}{\partial r_j}\right) \delta_{ik} D + r_j \delta_{ik} D' \left(\frac{\partial r}{\partial r_j}\right) = 0.
\end{aligned}$$

There are eight terms in this equation. Let us calculate one after another by using the tensor identities.

$$\left(\frac{\partial r_i}{\partial r_j}\right) r_j r_k A = \delta_{ij} r_j r_k A = \delta_{ii} r_i r_k A = r_i r_k A$$

$$r_i \left(\frac{\partial r_j}{\partial r_j}\right) r_k A = r_i \delta_{jj} r_k A = 3 r_i r_k A$$

$$r_i r_j \left(\frac{\partial r_k}{\partial r_j}\right) A = r_i r_j \delta_{kj} A = r_i r_k \delta_{kk} A = r_i r_k A$$

$$r_i r_j r_k \left(\frac{\partial r}{\partial r_j}\right) A' = r_i r_j r_k A' r_j / r = r_i r_k (r A')$$

$$\begin{aligned}
\left\{\left(\frac{\partial r_i}{\partial r_j}\right) \delta_{kj} + \left(\frac{\partial r_k}{\partial r_j}\right) \delta_{ij}\right\} B &= \left(\delta_{ij} \delta_{kj} + \delta_{kj} \delta_{ij}\right) B \\
&= (\delta_{ik} \delta_{kk} + \delta_{kk} \delta_{ik}) B = 2 \delta_{ik} B
\end{aligned}$$

$$\begin{aligned}
\left(r_i \delta_{kj} + r_k \delta_{ij}\right) \left(\frac{\partial r}{\partial r_j}\right) B' &= \left(r_i r_j \delta_{kj} + r_j r_k \delta_{ij}\right) (B'/r) \\
&= (r_i r_k \delta_{kk} + r_i r_k \delta_{ii}) (B'/r) = 2 r_i r_k (B'/r)
\end{aligned}$$

$$\left(\frac{\partial r_j}{\partial r_j}\right) \delta_{ik} D + r_j \delta_{ik} D' \left(\frac{\partial r}{\partial r_j}\right) = \delta_{jj} \delta_{ik} D + r_j r_j \delta_{ki} (D'/r) = 3 \delta_{ik} D + \delta_{ik} (r D').$$

Collecting this information and grouping as the coefficients of $r_i r_k$ and δ_{ik}, we can write the above equation in a simple form as

$$\left(r A' + 5A + \frac{2B'}{r}\right) r_i r_k + \left(2B + r D' + 3D\right) \delta_{ik} = 0$$

Since this equation holds for all values of i and k, we therefore, next equate the coefficient of $r_i r_k$ and δ_{ik} to zero, and this yields

$$r^2 A' + 5rA + 2B' = 0$$
$$2B + rD' + 3D = 0.$$

This is the required solution of the problem. The unknown variables are obtained in the previous section from the physical consideration of the problem. Here we have shown only how the powerful tensor calculus plays an important role.

6.9 Spectral theory of homogeneous turbulence

The energy spectrum tensor

We have already introduced the correlation tensor, $R_{ij}(\mathbf{r})$. We now introduce its three-dimensional Fourier transform, $\Phi_{ij}(\mathbf{K})$, defined by the relationship

$$R_{ij}(\mathbf{r}) = \int_{-\infty}^{\infty} \Phi_{ij}(\mathbf{K}) e^{i\mathbf{K}\cdot\mathbf{r}} d\mathbf{K}. \tag{6.85}$$

In this equation the integral is a triple integral taken over all values of each of three components of wavenumber space (K_1, K_2, K_3), $d\mathbf{K}$ being shorthand for the product $dK_1 dK_2 dK_3$. Then inverse of (6.85) is

$$\Phi_{ij}(\mathbf{K}) = \frac{1}{8\pi^3} \int_{-\infty}^{\infty} R_{ij}(\mathbf{r}) e^{-i\mathbf{K}\cdot\mathbf{r}} d\mathbf{r}. \tag{6.86}$$

It is worth noting that

$$R_{ij}(\mathbf{r}) = R_{ji}(-\mathbf{r}),$$

and an analogous property holds for $\Phi_{ij}(\mathbf{K})$, i.e.,

$$\Phi_{ij}(\mathbf{K}) = \Phi_{ji}(-\mathbf{K}).$$

We note, setting $\mathbf{r} = 0$ in (6.85), that

$$R_{ij}(0) = \overline{v_i(\mathbf{x}) v_j(\mathbf{x})} = \int_{-\infty}^{\infty} \Phi_{ij}(\mathbf{K}) d\mathbf{K}, \tag{6.87}$$

so $\Phi_{ij}(\mathbf{K})$ represents a density, in wavenumber space, of contributions to $\overline{v_i(\mathbf{x}) v_j(\mathbf{x})}$. By letting $i = j$ in (6.87), it is found that

$$\overline{v_i^2} = \int_{-\infty}^{\infty} \Phi_{ii}(\mathbf{K}) d\mathbf{K},$$

which has the dimension of energy per unit mass. For these reasons, $\Phi_{ij}(\mathbf{K})$ is called the energy spectrum tensor.

Note that the velocity covariance tensor, R_{ij} is defined as

$$\overline{v_i(\mathbf{x})v_j(\mathbf{x}')} = R_{ij}(\mathbf{x}' - \mathbf{x}),$$

$$\text{or} \qquad \overline{v_i v_j'} = R_{ij}(\mathbf{r})$$

Spectra in isotropic turbulence

In much the same way that the condition of isotropy determined the functional form of R_{ij}, it likewise follows that

$$\Phi_{ij}(\mathbf{K}) = B(K)\delta_{ij} + C(K)K_iK_j \qquad (6.88)$$

for some functions $B(K)$, $C(K)$. The equation of continuity then gives a relationship between B and C. From (6.85), since

$$\overline{v_i v_j'} = R_{ij}(\mathbf{r}) = \int_{-\infty}^{\infty} \Phi_{ij}(\mathbf{K}) \exp[i\mathbf{K} \cdot (\mathbf{x}' - \mathbf{x})]d\mathbf{K},$$

we have

$$0 = \overline{\frac{\partial v_i}{\partial x_i} v_j'} = -i \int_{-\infty}^{\infty} K_i \Phi_{ij}(\mathbf{K}) \exp[i\mathbf{K} \cdot (\mathbf{x}' - \mathbf{x})]d\mathbf{K},$$

and hence

$$K_i \Phi_{ij}(\mathbf{K}) = 0.$$

Upon substituting from (6.88) it is readily deduced that

$$B(K) + K^2 C(K) = 0.$$

With this relationship, (6.88) may be written

$$\Phi_{ij}(\mathbf{K}) = C(K)[K_iK_j - K^2\delta_{ij}]. \qquad (6.89)$$

It is now convenient at this point to introduce a new scalar function, defined by

$$\begin{aligned} E(K) &= \frac{\text{Energy contained in modes with scalar wavenumber } K \pm \frac{1}{2}dK}{dK} \\ &= \frac{\frac{1}{2}\Phi_{ii}(4\pi K^2 dK)}{dK} \\ &= 2\pi K^2 \Phi_{ii} \\ &= -4\pi K^4 C(K). \end{aligned}$$

Using $E(K)$ instead of $C(K)$, (6.89) finally becomes

$$\Phi_{ij}(\mathbf{K}) = \frac{E(K)}{4\pi K^4}[K^2\delta_{ij} - K_iK_j]. \qquad (6.90)$$

From experimental view point, it is usual to measure a one-dimensional spectrum function by the following definition

$$\phi(K_1) = \int_{-\infty}^{\infty}\int_{-\infty}^{\infty} \Phi_{11}(K_1, K_2, K_3) dK_2 dK_3.$$

We express this in terms of $E(K)$ by changing in effect from Cartesian coordinates (K_2, K_3) to polar coordinates. Thus, considering the $K_2 - K_3$ wave number plane in two-dimension (analogous to $x - y$ plane), we have

$$K_2 = K\cos\theta$$

$$K_3 = K\sin\theta$$

such that $K^2 = K_2^2 + K_3^2$, and $\theta = \tan^{-1}(\frac{K_3}{K_2})$, hence the elementary area

$$dK_2 dK_3 = (Kd\theta)(dK)$$

then the above integral can be written as

$$\begin{aligned}
\phi(K_1) &= \int_{-\infty}^{\infty}\int_{0}^{2\pi} \Phi_{11}(K_1, K, \theta) K dK d\theta \\
&= 2\pi \int_{K=K_1}^{\infty} \Phi_{11}(K_1, K) K dK \\
&= \frac{1}{2}\int_{K=K_1}^{\infty} \frac{E(K)}{K^3}[K^2 - K_1^2] dK.
\end{aligned}$$

This equation is easily inverted, by differentiating with respect to K_1. Thus we have

$$\phi'(K_1) = -K_1 \int_{K_1}^{\infty} \frac{E(K)}{K^3} dK$$

and $$E(K) = K_1^3 \frac{d}{dK_1}\left\{\frac{1}{K_1}\phi'(K_1)\right\}$$

Now we can write the expression for $E(K)$ as follows.

$$E(K) = K^2\phi''(K) - K\phi'(K). \tag{6.91}$$

The relationship between $E(K)$ and $f(r)$

It is convenient to use $\phi(K)$ as an intermediate function to obtain the relationship between $E(K)$ and $f(r)$. Thus upon substituting for Φ_{11} from (6.86)

$$\begin{aligned}
\phi(K_1) &= \frac{1}{8\pi^3}\int_{-\infty}^{\infty} R_{11}(r_1, r_2, r_3)\exp[-i(K_1 r_1 + K_2 r_2 + K_3 r_3)] \\
&\quad dr_1 dr_2 dr_3 dK_2 dK_3 \\
&= \frac{1}{2\pi}\int_{-\infty}^{\infty} R_{11}(r_1, 0, 0)\exp[-iK_1 r_1] dr_1,
\end{aligned}$$

after twice using the result that

$$\int_{-\infty}^{\infty}\int_{-\infty}^{\infty} F(x)e^{-ixy}dxdy = 2\pi F(0).$$

Substituting for $R_{11}(r,0,0) = u^2 f(r)$ where $u^2 = \overline{v_1^2} = \overline{v_2^2} = \overline{v_3^2}$, and $f(r)$ is known as longitudinal velocity correlation coefficient, we obtain

$$\begin{aligned}\phi(K) &= \frac{u^2}{2\pi}\int_{-\infty}^{\infty} f(r)e^{-iKr}dr \\ &= \frac{u^2}{\pi}\int_0^{\infty} \cos Krf(r)dr. \end{aligned} \tag{6.92}$$

The value of $E(K)$ then follows from (6.91), and is given by

$$E(K) = \frac{u^2}{\pi}\int_0^{\infty} f(r)\left\{Kr\sin Kr - K^2r^2\cos Kr\right\}dr. \tag{6.93}$$

To invert this equation we note that (6.92) leads to

$$\begin{aligned}u^2 f(r) &= 2\int_0^{\infty} \phi(K_1)\cos K_1 r dK_1 \\ &= \int_0^{\infty} \cos K_1 r dK_1 \int_{K_1}^{\infty} \frac{E(K)}{K^3}\left[K^2 - K_1^2\right]dK.\end{aligned}$$

We now invert the order of integration. Hence we obtain

$$\begin{aligned}u^2 f(r) &= \int_0^{\infty} \frac{E(K)}{K^3}dK\int_0^{K}\left[K^2 - K_1^2\right]\cos K_1 r dK_1 \\ &= 2\int_0^{\infty} E(K)\left(\frac{\sin Kr}{K^3r^3} - \frac{\cos Kr}{K^2r^2}\right)dK. \end{aligned} \tag{6.94}$$

Equations (6.93) and (6.94) have a number of interesting consequences. We deduce from (6.94) that

$$\begin{aligned}u^2\ell &= u^2\int_0^{\infty} f(r)dr \\ &= 2\int_0^{\infty} E(K)dK\int_0^{\infty}\left(\frac{\sin Kr}{K^3r^3} - \frac{\cos Kr}{K^2r^2}\right)dr \\ &= 2\int_0^{\infty}\frac{E(K)}{K}dK\int_0^{\infty}\left(\frac{\sin t}{t^3} - \frac{\cos t}{t^2}\right)dt \\ &= \frac{\pi}{2}\int_0^{\infty}\frac{E(K)}{K}dK.\end{aligned}$$

But we know that

$$\frac{3}{2}u^2 = \int_0^\infty E(K)dK,$$

and hence we have

$$\ell = \frac{3\pi}{4}\frac{\int_0^\infty \frac{E(K)}{K}dK}{\int_0^\infty E(K)dK}$$

$$= \text{mean value of } \frac{3\pi}{4}K^{-1},$$

the mean value being taken with respect to the energy spectrum. The result verifies that ℓ is representative of the size of the energy-bearing eddies.

We now expand $\sin Kr$ and $\cos Kr$ in powers of Kr, and so (6.94) becomes

$$u^2 f(r) = \frac{2}{3}\int_0^\infty E(K)\left\{1 - \frac{1}{10}K^2r^2 + \cdots\right\}dK.$$

so that

$$\frac{u^2}{\lambda^2} = -u^2 f''(0) = \frac{2}{15}\int_0^\infty K^2E(K)dK.$$

Accordingly the rate of energy dissipation per unit mass is given by

$$\epsilon = -\frac{d}{dt}\left(\frac{3}{2}u^2\right) = \frac{15\nu u^2}{\lambda^2}$$

$$= 2\nu\int_0^\infty K^2E(K)dK, \tag{6.95}$$

which depends upon the values of $E(K)$ for large K, that is upon the small eddies.

Rate of change of the energy spectrum

We write the equation already deduced in the last section,

$$\frac{\partial}{\partial t}(R_{ij}) = T_{ij} + P_{ij} + 2\nu\nabla^2 R_{ij}$$

and take the three-dimensional Fourier transform of each term. This transformation leads to

$$\frac{\partial}{\partial t}(\Phi_{ij}) = \Gamma_{ij} + \Lambda_{ij} - 2\nu K^2\Phi_{ij}, \tag{6.96}$$

where

$$\Gamma_{ij} = \frac{1}{8\pi^3}\int T_{ij}e^{-i\mathbf{K}\cdot\mathbf{r}}d\mathbf{r}.$$

and

$$\Lambda_{ij} = \frac{1}{8\pi^3} \int P_{ij} e^{-i\mathbf{K}\cdot\mathbf{r}} d\mathbf{r}.$$

It is worth noting here that for isotropic turbulence $P_{ij} = 0$ and $\Lambda_{ij} = 0$. For non-isotropic homogeneous turbulence, it was shown that $P_{ii} = 0$ none the less, and hence $\Lambda_{ii} = 0$. When we let $i = j$ in (6.96) we therefore obtain the following result

$$\frac{\partial}{\partial t}(\Phi_{ii}) = \Gamma_{ii} - 2\nu K^2 \Phi_{ii},$$

the absence of the pressure term indicates that these do not affect the total energy contributed by any small region in wave-number space, but merely change the directional distribution of this energy.

Turning now to isotropic turbulence, we multiply by $2\pi K^2$ and write

$$E(K) = 2\pi K^2 \Phi_{ii}$$

so that

$$\frac{\partial}{\partial t}(E(K)) = T(K) - 2\nu K^2 E(K),$$

where

$$T(K) = 2\pi K^2 \Gamma_{ii}.$$

After formal integration this becomes

$$\frac{\partial}{\partial t}\left\{\int_0^K E(K_1)dK_1\right\} = S(K) - 2\nu \int_0^K K_1^2 E(K_1)dK_1, \tag{6.97}$$

with

$$S(K) = 2\pi \int_0^K K_1^2 \Gamma_{ii}(K_1)dK_1.$$

This equation has a simple physical interpretation that the energy of the large eddies changes because of a transfer $S(K) < 0$ from the small eddies, and because of viscous dissipation.

In an attempt to solve the energy decay equation, Heisenberg has made certain assumptions which are likely to be valid for the energy-bearing eddies, by suggesting that the extraction of energy from the large eddies takes place as if an eddy-viscosity caused an increased dissipation in the larger eddies. Thus (6.97) is written as

$$\frac{\partial}{\partial t}\left\{\int_0^K E(K_1)dK_1\right\} = -2(\nu_e + \nu) \int_0^K K_1^2 E(K_1)dK_1. \tag{6.98}$$

By further assuming that $\nu_e(K)$ shall be independent of ν (a reasonable suggestion for those eddies outside the dissipation range), and that $\nu_e(K)$ depends upon $E(K_1)$ for all $K_1 > K$, that is on all the smaller eddies, then dimensional reasoning suggests that

$$\nu_e(K) \propto \int_K^\infty (K_1 E(K_1))^{1/2} \frac{1}{K_1} \frac{dK_1}{K_1},$$

where the three factors in the integrand have the dimensions of velocity, length and a number, respectively. Thus, we write

$$\nu_e(K) = \gamma \int_K^\infty (E(K_1))^{\frac{1}{2}} K_1^{-3/2} dK_1,$$

and (6.98) then becomes

$$\frac{\partial}{\partial t}\left\{\int_0^K E(K_1)dK_1\right\}$$
$$= -2\left\{\nu + \gamma \int_0^\infty (E(K_1))^{\frac{1}{2}} K_1^{-3/2} dK_1\right\} \int_0^K K_1^2 E(K_1)dK_1.$$

This equation must be solved numerically to give $E(K)$. A fair test of its accuracy would be its prediction of $f(r)$, which is determined by the medium-sized and larger eddies. It is found that if γ is taken to be approximately 0.45, the calculated $f(r)$ at a Reynolds number of 240 is in good agreement with experimental results.

6.10 Probability distribution of $u(\mathbf{x})$

Consider a given flow, in which a velocity component, say v_1, is measured at a fixed point $\mathbf{x}$. We have already seen that in a truly turbulent flow the value of $v_1(\mathbf{x})$ will vary randomly. Numerous experimental measurements have been made of the probability distribution of such a single velocity component in approximately isotropic turbulence, and it has been found to have normal or Gaussian distribution. The accuracy to which this is true is considerable. For example, values of the flatness factor $\overline{v_1^4}/[\overline{(v_1^2)}]^2$ have been measured, such values having biased in favour of large values of $|v_1|$. The experimental results yield flatness factors lying between 2.9 and 3.0 as compared with the value 3.0 appropriate to a normal distribution. Further, the skewness factor $\overline{v_1^3}/[\overline{(v_1^2)}]^{3/2}$ has been experimentally found to be very close to zero. These results are not hard to understand. The velocity at any point is subject to the influence of a large number of random eddies in its neighborhood, so the resultant normal probability distribution is not surprising.

Let us now consider velocity components at two different fixed points, and we want to know the probability distribution of $v_1(\mathbf{x}) - v_1(\mathbf{x}')$. We may surmise that if $r = |\mathbf{x} - \mathbf{x}'|$ is sufficiently large, the two velocity components will be independent, and hence $v_1 - v_1'$ will have a normal distribution. On the other hand when r is sufficiently small the values of v_1 and v_1' will be well correlated, and the probability and the probability distribution will be determined by the source of the correlation, that is, by the equation of motion. The experimental results bear out these contentions. For example, values of flatness factor

$$\overline{(v_1 - v_1')^4}/[\overline{(v_1 - v_1')^2}]^2$$

have been measured for various values of separation of the two points. It is found that it deceases from roughly 3.6 when $r = 0$ to roughly 2.9 when r is large. By comparison with the correlation function $f(r)$ it is found that the flatness factor lies between 2.9 and 3.1 provided r is sufficiently large that $f(r) < 0.7$. In other words, the relationship between the fourth-order mean value $\overline{(v_1 - v_1')^4}$ and the second-order mean value $\overline{[(v_1 - v_1')^2]}$ is that corresponding to a normal joint probability distribution at least as far as the larger eddies are concerned. On the other hand, measured values of the skewness factor

$$\overline{(v_1 - v_1')^3}/[\overline{(v_1 - v_1')^2}]^{3/2}$$

fall to zero only for exceedingly large r, but not for values of r for which $f(r)$ is significantly non-zero. In other words, the relationship between the third-order and second-order mean values is not that appropriate to a normal distribution.

The above results lead to a reasonable working hypothesis, which has proved very useful, the part of the probability distribution of $v(\mathbf{x})$ which is determined by the larger eddies is approximately normal, at least as far as the values of the velocity at no more than two points are concerned, and particularly as regards the relation between even-order two-point mean values . This enables us to express complicated velocity product mean values in terms of the basic correlation function $f(r)$.

6.11 Calculation of the pressure covariance in isotropic turbulence

In this section we shall discuss an illustration and application of the ideas presented in the previous section. Here we shall now briefly indicate how the pressure covariance $\overline{p(\mathbf{x})p(\mathbf{x}')} = \overline{pp'}$ may be expressed in terms of $f(r)$. We begin with the equation

$$\frac{\partial v_i}{\partial t} + \frac{\partial}{\partial x_j}(v_i v_j) = -\frac{1}{\rho}\frac{\partial p}{\partial x_i} + \nu \nabla^2 v_i,$$

which we differentiate with respect to x_i. Then by virtue of the continuity equation $\frac{\partial v_i}{\partial x_i} = 0$ we have

$$\frac{1}{\rho}\frac{\partial^2 p}{\partial x_i^2} = \frac{\partial^2}{\partial x_i \partial x_j}(v_i v_j)$$

Likewise

$$\frac{1}{\rho}\frac{\partial^2 p'}{\partial x_k'^2} = \frac{\partial^2}{\partial x_k' \partial x_l'}(v_k' v_l')$$

and multiplication leads to the result

$$\frac{1}{\rho^2}\frac{\partial^4}{\partial x_i^2 \partial x_k'^2}(pp') = \frac{\partial^4}{\partial x_i \partial x_j \partial x_k' \partial x_l'}(v_i v_j v_k' v_l')$$

We take mean value and write

$$\frac{1}{\rho^2}\overline{pp'} = P(r),$$

$$\frac{\overline{\partial^4 v_i v_j v_k' v_l'}}{\partial x_i \partial x_j \partial x_k' \partial x_l'} = \frac{\overline{\partial^4 v_i v_j v_k' v_l'}}{\partial r_i \partial r_j \partial r_k \partial r_l} = W(r). \tag{6.99}$$

Then, since $\frac{\partial^2}{\partial x_i^2} \equiv \nabla^2$ becomes $\frac{1}{r^2}\frac{d}{dr}\left(r^2 \frac{d}{dr}\right)$ for function of r in cylindrical polar coordinates, we have

$$\frac{1}{r^2}\frac{d}{dr}\left[r^2 \frac{d}{dr}\left\{\frac{1}{r^2}\frac{d}{dr}\left(r^2 \frac{dP}{dr}\right)\right\}\right] = W(r),$$

which may be integrated successively four times to yield

$$P(r) = \frac{1}{6r}\int_r^\infty y(y-r)^3 W(y)dy. \tag{6.100}$$

On the basis of the ideas presented above and taking the approximations, we may express $W(r)$ in terms of $f(r)$ as follows. When the joint probability distribution of $\mathbf{v}$ and $\mathbf{v}'$ is normal it may be shown that

$$\overline{v_i v_j v_k' v_l'} = (\overline{v_i v_j})(\overline{v_k' v_l'}) + (\overline{v_i v_k'})(\overline{v_j v_l'}) + (\overline{v_i v_l'})(\overline{v_i v_k'}).$$

The proof of this result can be found in Batchelor's *The Theory of Homogeneous Turbulence* [1]. In this equation we note that the first term on the right-hand side is a constant. Thus, from (6.99), we have

$$\begin{aligned} W(r) &= \frac{\partial^4}{\partial r_i \partial r_j \partial r_k \partial r_l}\left((\overline{v_i v_k'})(\overline{v_j v_l'}) + (\overline{v_i v_l'})(\overline{v_j v_k'})\right) \\ &= 2\frac{\partial^2(\overline{v_i v_k'})}{\partial r_j \partial r_l} \cdot \frac{\partial^2(\overline{v_j v_l'})}{\partial r_i \partial r_k}, \end{aligned} \tag{6.101}$$

since all the remaining terms which arise in successive differentiation of a product are zero by virtue of the continuity equation.

We know that

$$R_{ij}(\mathbf{r}) = u^2\left\{\left(f + \frac{1}{2}rf'\right)\delta_{ij} - \frac{1}{2r}f' r_i r_j\right\}$$

and hence by this equation we have

$$\overline{v_i v_k'} = u^2 \left\{ \left(f + \frac{1}{2} r f' \right) \delta_{ik} - \frac{1}{2r} f' r_i r_k \right\},$$

and upon substitution into (6.101) it may be shown, after much tedious but straightforward algebraic reduction, that

$$W(r) = 4u^2 \left\{ 2(f'')^2 + 2f' f'' + \frac{10}{r} f' f'' + \frac{3}{r^2} (f')^2 \right\}.$$

We then substitute into (6.100), and after integration by 'parts' several times it may be shown that

$$P(r) = \frac{1}{\rho^2} (\overline{pp'}) = 2u^4 \int_r^\infty \left(y - \frac{r^2}{y} \right) [f'(y)]^2 dy.$$

Thus, once f is prescribed, $\overline{PP'}$ follows by integration. We here note that, by setting $r = 0$ we have that

$$\overline{p^2} = 2\rho^2 u^4 \int_0^\infty y (f')^2 dy,$$

and with the values of f this integral yields (see the previous section)

$$\overline{p^2} = 0.34 \rho^2 u^4.$$

Example 6.6

Show that

$$\int_{-\infty}^{\infty} \int_{-\infty}^{\infty} F(x) e^{\pm ixy} dx dy = 2\pi F(0).$$

Proof

Let us denote the integral as I_+ with positive exponential such that

$$\begin{aligned} I_+ &= \int_{-\infty}^{\infty} \int_{-\infty}^{\infty} F(x) e^{ixy} dx dy \\ &= \int_{-\infty}^{\infty} F(x) \left(\int_{-\infty}^{\infty} e^{ixy} dy \right) dx \end{aligned}$$

$$= \int_{-\infty}^{\infty} F(x) \left[\frac{e^{ixy}}{ix} \right]_{y=-\infty}^{\infty} dx$$

$$= \int_{-\infty}^{\infty} F(x) \lim_{y \to \infty} \left(\frac{e^{ixy} - e^{-ixy}}{ix} \right)$$

$$= \int_{-\infty}^{\infty} F(x) \lim_{y \to \infty} 2\pi \left\{ \frac{\sin(xy)}{\pi x} \right\} dx$$

$$= 2\pi \int_{-\infty}^{\infty} F(x)\delta(x) dx$$

$$= 2\pi F(0).$$

Note that $\lim_{y \to \infty} \frac{\sin(xy)}{x} = \delta(x)$ by definition. Similar analysis can be made taking the negative exponential and the result will be identical. Let us show it explicitly.

$$I_- = \int_{-\infty}^{\infty} \int_{-\infty}^{\infty} F(x) e^{-ixy} dx dy$$

$$= \int_{-\infty}^{\infty} F(x) \left(\int_{-\infty}^{\infty} e^{-ixy} dy \right) dx$$

$$= \int_{-\infty}^{\infty} F(x) \left[\frac{e^{-ixy}}{-ix} \right]_{y=-\infty}^{\infty} dx$$

$$= \int_{-\infty}^{\infty} F(x) \lim_{y \to \infty} \left(\frac{e^{ixy} - e^{-ixy}}{ix} \right)$$

$$= \int_{-\infty}^{\infty} F(x) \lim_{y \to \infty} 2\pi \left\{ \frac{\sin(xy)}{\pi x} \right\} dx$$

$$= 2\pi \int_{-\infty}^{\infty} F(x)\delta(x) dx$$

$$= 2\pi F(0).$$

Hence $I_+ = I_- = I$. This is the required proof.

Example 6.7

Show that

$$\int_0^{\infty} \left(\frac{\sin t}{t^3} - \frac{\cos t}{t^2} \right) dt = \frac{\pi}{4}.$$

Proof

Let us consider the left-hand side as I.

$$
\begin{aligned}
I &= \int_0^\infty \left(\frac{\sin t}{t^3} - \frac{\cos t}{t^2} \right) dt \\
&= \left[\frac{\sin t}{-2t^2} \right]_0^\infty - \int_0^\infty \left\{ \frac{\cos t}{-2t^2} + \frac{\cos t}{t^2} \right\} dt \\
&= \left[\frac{\sin t}{-2t^2} \right]_0^\infty - \int_0^\infty \left\{ \frac{\cos t}{2t^2} \right\} dt \\
&= \lim_{t\to 0} \left(\frac{1}{2t} \right) - \left[\frac{\cos t}{-2t} \right]_0^\infty + \int_0^\infty \frac{\sin t}{2t} dt \\
&= \lim_{t\to 0} \left(\frac{1}{2t} - \frac{1}{2t} \right) + \frac{1}{2} \int_0^\infty \frac{\sin t}{t} dt \\
&= \frac{\pi}{4}.
\end{aligned}
$$

Hence the proof. Note that $\int_0^\infty \frac{\sin t}{t} dt = \frac{\pi}{2}$.

Example 6.8

The Fourier transform in three-dimensions is defined as

$$
\begin{aligned}
&\Phi_{11}(K_1, K_2, K_3) \\
&= \frac{1}{8\pi^3} \int_{-\infty}^{\infty} \int_{-\infty}^{\infty} \int_{-\infty}^{\infty} R_{11}(r_1, r_2, r_3) e^{-i(K_1 r_1 + K_2 r_2 + K_3 r_3)} dr_1 dr_2 dr_3.
\end{aligned}
$$

Here Φ_{11} is called the spectrum and R_{11} is the correlation coefficient. Now let us define for our convenience

$$
\phi(K_1) = \int_{-\infty}^{\infty} \int_{-\infty}^{\infty} \Phi_{11}(K_1, K_2, K3) dK_2 dK_3.
$$

Then show that

$$
\phi(K_1) = \frac{1}{2\pi} \int_{-\infty}^{\infty} R_{11}(r_1, 0, 0) e^{-iK_1 r_1} dr_1.
$$

Proof

Given that

$$\begin{aligned}
\phi(K_1) &= \int_{-\infty}^{\infty}\int_{-\infty}^{\infty} \Phi_{11}(K_1, K_2, K3)dK_2dK_3 \\
&= \int_{-\infty}^{\infty}\int_{-\infty}^{\infty}\left\{\frac{1}{8\pi^3}\int_{-\infty}^{\infty}\int_{-\infty}^{\infty}\int_{-\infty}^{\infty} R_{11}(r_1, r_2, r_3)\right\} \\
&\quad e^{-i(K_1r_1+K_2r_2+K_3r_3)}dr_1dr_2dr_3dK_2dK_3 \\
&= \frac{1}{8\pi^3}\left(\int_{r_1=-\infty}^{\infty} e^{-iK_1r_1}dr_1\right)\left(\int_{r_2=-\infty}^{\infty}\int_{K_2=-\infty}^{\infty} e^{-iK_2r_2}dK_2dr_2\right) \\
&\quad \left(\int_{r_3=-\infty}^{\infty}\int_{K_3=-\infty}^{\infty} R_{11}(r_1, r_2, r_3)e^{-iK_3r_3}dK_3dr_3\right) \\
&= \frac{1}{4\pi^2}\left(\int_{r_1=-\infty}^{\infty} e^{-iK_1r_1}dr_1\right) \\
&\quad \left(\int_{r_2=-\infty}^{\infty}\int_{K_2=-\infty}^{\infty} R_{11}(r_1, r_2, 0)e^{-iK_2r_2}dK_2dr_2\right) \\
&= \frac{1}{2\pi}\int_{r_1=-\infty}^{\infty} R_{11}(r_1, 0, 0)e^{-iK_1r_1}dr_1.
\end{aligned}$$

This is the required proof. In obtaining this result we have used the result of Example 11.6.

Example 6.9

Fluid flows with uniform speed U_0 in the region $y > 0, x < 0$. When $x > 0$ it mixes with fluid at rest in the region $y < 0$. Show that the equation for the mean velocity in the turbulent mixing region is, on the basis of mixing-length theory,

$$\begin{aligned}
u\frac{\partial u}{\partial x} + v\frac{\partial u}{\partial y} &= 2\ell^2\frac{\partial u}{\partial y}\frac{\partial^2 u}{\partial y^2}, \\
\frac{\partial u}{\partial x} + \frac{\partial v}{\partial y} &= 0,
\end{aligned}$$

where the mixing-length $\ell = ax^m$ is proportional to the width of the mixing region.

Then look for a solution of the form

$$u = \frac{\partial \psi}{\partial y}, \quad v = -\frac{\partial \psi}{\partial x}$$

where ψ is a stream function which satisfies the continuity equation. Consider a similar solution in the form $\psi(x, y) = x^n f(\eta)$ such that $\eta = y/x^m$ known as the similarity variable. Deduce that for $m = 1$, the equation for $f(\eta)$ is the following ordinary differential equation

$$2a^2 f'' f''' + nff'' = (n-1)f'^2.$$

By consideration of the boundary condition on u, deduce that $n = 1$, and thus that

$$2a^2 f''' + f = 0.$$

Solution

The two-dimensional turbulent boundary layer equations are given by

$$u\frac{\partial u}{\partial x} + v\frac{\partial u}{\partial y} = U\frac{dU}{dx} + \frac{1}{\rho}\frac{\partial \tau}{\partial y},$$
$$\frac{\partial u}{\partial x} + \frac{\partial v}{\partial y} = 0,$$

where $\tau = \mu\frac{\partial u}{\partial y} - \rho\overline{u'v'}$. From experimental observation, Prandtl's mixing-length theory can be derived as $-\rho\overline{u'v'} = \ell^2\frac{\partial u}{\partial y}|\frac{\partial u}{\partial y}| = \ell^2\left(\frac{\partial u}{\partial y}\right)^2$. Thus, we have $\tau = \mu\frac{\partial u}{\partial y} + \ell^2\left(\frac{\partial u}{\partial y}\right)^2$. Hence

$$\frac{\partial \tau}{\partial y} = \mu\frac{\partial^2 u}{\partial y^2} + 2\ell^2\left(\frac{\partial u}{\partial y}\right)\left(\frac{\partial^2 u}{\partial y^2}\right)$$
$$= 2\ell^2\left(\frac{\partial u}{\partial y}\right)\left(\frac{\partial^2 u}{\partial y^2}\right)$$

Here the viscous stress $\mu\frac{\partial^2 u}{\partial y^2} = 0$ because it is very small in comparison with the Reynolds stress. Also we have $\frac{dU}{dx} = 0$ because U is a uniform stream speed. Thus collecting all these information, we can write the boundary layer equations in the turbulent mixing region as

$$u\frac{\partial u}{\partial x} + v\frac{\partial u}{\partial y} = 2\ell^2\frac{\partial u}{\partial y}\frac{\partial^2 u}{\partial y^2},$$
$$\frac{\partial u}{\partial x} + \frac{\partial v}{\partial y} = 0,$$

This set of equations can be solved by the similarity technique. We introduce a stream function $\psi(x, y)$ which satisfies the continuity equation. This function is related to the velocity components $u = \frac{\partial \psi}{\partial y}$ and $v = -\frac{\partial \psi}{\partial x}$. We look for a solution in the form $\psi(x, y) = x^n f(\eta)$ where $\eta = y/x^m$ is defined as the similarity variable.

With these information, we deduce the different terms in the above equations in terms of $f(\eta)$. The following calculations are presented here.

$$u = \frac{\psi}{\partial y} = x^{n-m} f'$$

$$v = -\frac{\partial \psi}{\partial x} = -nx^{n-1} f + mx^{n-1}\eta f'$$

$$\frac{\partial u}{\partial y} = x^{n-2m} f''$$

$$\frac{\partial^2 u}{\partial y^2} = x^{n-3m} f'''$$

$$\frac{\partial u}{\partial x} = (n-m)x^{n-m-1} f' - mx^{n-m-1}\eta f''$$

Substituting these values into the above boundary layer equation yields

$$\begin{aligned} x^{2n-2m-1}(n-m)f'^2 &- mx^{2n-2m-1}\eta f' f'' \\ -nx^{2n-2m-1} ff'' &+ mx^{2n-2m-1}\eta f' f'' \\ &= 2\ell^2 x^{2n-5m} f'' f''' \end{aligned}$$

which can be written after reduction as

$$x^{2n-2m-1}(n-m)f'^2 - nx^{2n-2m-1} ff'' = 2\ell^2 x^{2n-5m} f'' f'''$$

Now if we let $\ell = ax^m$ such that $\ell^2 = a^2 x^{2m}$, then the above equation reduces to

$$(n-m)f'^2 - nff'' = 2a^2 x^{1-m} f'' f'''.$$

Let $m = 1$ in which case the x variable will be eliminated from the differential equation and the differential equation will be completely in terms of the similarity variable η. This is a reasonable assumption so far as the boundary conditions are concerned. Hence the differential equation becomes

$$2a^2 f'' f''' + nff'' = (n-1)f'^2.$$

If we let $n = 1$, then the ordinary differential equation becomes simply

$$2a^2 f''' + f = 0.$$

This is the required result.

Example 6.10

Starting from the kinetic-energy integral equation for a turbulent boundary layer,

$$\frac{d}{dx}(\rho U^3 \delta_3) = 2\int_0^\infty \tau \frac{\partial u}{\partial y} dy,$$

and assuming that the turbulent shearing stress is approximately

$$\tau \approx \tau_w + y\frac{dp}{dx},$$

show that

$$\frac{d}{dx}(\rho U^3 \delta_3) \approx 2U\left(\tau_w + \delta_1\frac{dp}{dx}\right).$$

Solution

The kinetic-energy integral equation for a turbulent boundary layer is given by

$$\frac{d}{dx}(\rho U^3 \delta_3) = 2\int_0^\infty \tau\frac{\partial u}{\partial y}dy.$$

We know that the turbulent shearing stress is defined as $\tau = \tau_w + \frac{dp}{dx}y$. Thus substituting the value of τ in the above equation, yields

$$\begin{aligned}
\frac{d}{dx}(\rho U^3 \delta_3) &= 2\int_0^\infty \left(\tau_w + \frac{dp}{dx}y\right)\frac{\partial u}{\partial y}dy \\
&= 2\left\{\int_0^\infty \tau_w\left(\frac{\partial u}{\partial y}\right)dy + \int_0^\infty \frac{dp}{dx}y\frac{\partial u}{\partial y}dy\right\} \\
&= 2U\left\{\int_0^\infty \tau_w\left(\frac{\partial(u/U)}{\partial y}\right)dy + \int_0^\infty \frac{dp}{dx}y\frac{\partial(u/U)}{\partial y}dy\right\} \\
&= 2U\left(\tau_w + \delta_1\frac{dp}{dx}\right).
\end{aligned}$$

We can simply show that $\int_0^\infty \tau_w(\frac{\partial(u/U)}{\partial y})dy = \tau_w$, and

$$\int_0^\infty \frac{dp}{dx}y\frac{\partial(u/U)}{\partial y}dy = \frac{dp}{dx}\int_0^\infty\left(1-\frac{u}{U}\right)dy = \frac{dp}{dx}\delta_1.$$

Thus the required result is

$$\frac{d}{dx}(\rho U^3 \delta_3) \approx 2U\left(\tau_w + \delta_1\frac{dp}{dx}\right).$$

Example 6.11

Show that in homogeneous isotropic turbulence, the rate of energy dissipation per unit mass is

$$\epsilon = \nu\left(\frac{\partial v_i}{\partial x_j}\right)^2 = \frac{15\nu u^2}{\lambda^2}.$$

Use a physical argument to deduce the approximate equation

$$\frac{d}{dt}(u^2) = -Au^3/\ell,$$

for the energy decay of grid turbulence. Hence find a relationship between λ and ℓ, and comment very briefly on the significance of these two length scales.

Solution

We start with the definitions of longitudinal velocity correlation coefficient, $f(r)$, defined by $R_{11}(r, 0, 0) = u^2 f(r)$ where $u^2 = \overline{v_1^2} = \overline{v_2^2} = \overline{v_3^2}$. This involves a correlation between the velocity components at two points, each component being parallel to the vector separation of the points. In actual practice,

$$R_{ij}(\mathbf{r}) = u^2 \left\{ \left(f + \frac{1}{2} r f' \right) \delta_{ij} - \frac{1}{2r} f' r_i r_j \right\}.$$

We now introduce the lateral velocity correlation coefficient, $g(r)$, defined by $R_{ii}(0, r, 0) = u^2 g(r) = u^2 (f + \frac{1}{2} r f')$, which involves the velocity components at two points, each component being normal to the vector separation of the points. We now introduce two scales of turbulence. The first is called the longitudinal integral scale, and is defined by

$$\ell = \int_0^\infty f(r) dr.$$

This is nothing but the representative of the energy-bearing eddies. To define the second scale we write

$$f(r) = f(0) + f'(0)r + \frac{1}{2} f''(0) r^2 + \cdots = f(0) + \frac{1}{2} f''(0) r^2 + \cdots$$

for small r. Since $f(r)$ is an even function, so $f'(0) = 0$, and the expansion is appropriate. We also note that $f''(0) < 0$, so we write $f''(0) = -1/\lambda^2$, and hence $f(r) = f(0) - \frac{r^2}{2\lambda^2} = 1 - \frac{r^2}{2\lambda^2}$ because the maximum value of $f(r) = f(0) = 1$. It may likewise be shown that when r is small the function $g(r)$ may be expanded as $g(r) = f(r) + \frac{1}{2} r f'(r) = 1 - \frac{r^2}{\lambda^2} + \cdots$. The rate of energy dissipation per unit mass may conveniently be expressed in terms of the length λ, as

$$\begin{aligned}\epsilon &= \nu\overline{\left(\frac{\partial v_i}{\partial x_j}\right)^2}\\ &= \nu\left\{3 \text{ terms like } \overline{\left(\frac{\partial v_1}{\partial x_1}\right)^2} + 6 \text{ terms like } \overline{\left(\frac{\partial v_1}{\partial x_2}\right)^2}\right\}\\ &= \nu\left\{3\frac{u^2}{\lambda^2} + 6\frac{2u^2}{\lambda^2}\right\}\\ &= \frac{15\nu u^2}{\lambda^2}.\end{aligned}$$

But we know that

$$\frac{\partial}{\partial t}\left(\frac{1}{2}v_i^2\right) = -\nu\left(\frac{\partial v_i}{\partial x_j}\right) = -\epsilon.$$

We drop the bar for simplicity. And hence $\frac{d}{dt}(\frac{3}{2}u^2) = -\epsilon = -\frac{15\nu u^2}{\lambda^2}$. Thus we have $\frac{d}{dt}(u^2) = -\frac{10\nu u^2}{\lambda^2}$.

We also know that in the case of homogeneous turbulence, the mean square vorticity is given by

$$\frac{\partial}{\partial t}\left(\frac{1}{2}v_i^2\right) = -\nu\omega_i^2 = -\nu\omega^2 = -\nu\frac{15u^2}{\lambda^2}.$$

And hence we have

$$\omega^2 = \frac{15u^2}{\lambda^2}.$$

Since ℓ is a characteristic length of these eddies, we have approximately

$$-\frac{d}{dt}\left(\frac{3}{2}u^2\right) \approx \frac{3}{2}u^2/(\ell/u) = \frac{3}{2}\frac{u^3}{\ell},$$

and hence

$$-\frac{d}{dt}(u^2) = \frac{10\nu u^2}{\lambda^2} \approx \frac{u^3}{\ell}.$$

This can be written as

$$\frac{d}{dt}(u^2) = -A\frac{u^3}{\ell},$$

where A is a constant. This is known as the energy decay for grid turbulence. Now we can relate λ and ℓ in the following manner.

$$-\frac{d}{dt}(u^2) = \frac{10\nu u^2}{\lambda^2} = A\frac{u^3}{\ell}.$$

Therefore

$$\lambda^2 = \frac{10\nu\ell}{Au}.$$

Thus the square of the viscous dissipation length is directly proportional to the longitudinal integral scale ℓ (representative of the energy bearing eddies) and inversely proportional to the mean eddy velocity.

From the experimental point of view, the sort of turbulence produced under laboratory conditions, when a stream of velocity U impinges on the grids of mesh M, it is found that

$$R_\ell = \frac{u\ell}{\nu} \approx 0.01 R_M = 0.01\frac{UM}{\nu}.$$

Hence we have from the above analysis that

$$\frac{\ell^2}{\lambda^2} \approx 0.1 R_\ell,$$

and it follows that

$$\frac{\ell^2}{\lambda^2} \approx 0.001 R_M.$$

Since typical R_M lies in a range from 2,000 to 100,000, it follows that $(\ell/\lambda)^2$ will lie in the range 2-100, and so (ℓ/λ) will be between 1 and 10.

For further information about this important topic, the interested reader is referred to the work of Lamb, H. [3] to Rahman, M. [7] as listed in the reference section.

6.12 Exercises

1. Define the velocity covariance tensor $R_{ij}(\mathbf{r})$ for homogeneous turbulence, and show that it satisfies a dynamical equation of the form

$$\frac{\partial R_{ij}}{\partial t} = T_{ij} + P_{ij} + 2\nu\nabla^2 R_{ij},$$

   where T_{ij} is the inertia tensor and P_{ij} the pressure tensor. Show that P_{ij} is zero when the turbulence is isotropic.

   Deduce the equation for the energy spectrum function $E(K)(= 2\pi K^2\Phi_{ii}(K))$ for homogeneous isotropic turbulence, and comment on the difficulties of solving it.

2. Given that the energy spectrum tensor for homogeneous isotropic turbulence is of the form

$$\Phi_{ij}(K) = \frac{E(K)}{4\pi K^4}\left(K^2\delta_{ij} - K_iK_j\right),$$

   derive the two equations

$$\phi(K_1) = \frac{1}{2}\int_{K_1}^{\infty} \frac{E(K)}{K^3}(K^2 - K_1^2)dK,$$

$$E(K_1) = K_1^3\frac{d}{dK_1}\left(\frac{1}{K_1}\frac{d\phi}{dK_1}\right)$$

relating the one-dimensional spectrum function $\phi(K_1)$ and the energy spectrum function $E(K)$. Infer the behaviours of ϕ and E for small values of K, given by

$$u^2\ell = \frac{\pi}{2}\int_0^\infty K^{-1}E(K)dK;$$

sketch ϕ and E as a function K.

3. Define the longitudinal velocity correlation $f(r)$, the lateral velocity correlation $g(r)$ and the energy dissipation length λ in homogeneous isotropic turbulence. Sketch typical curves of $f(r)$ and $g(r)$, and comment briefly on their respective shapes.

   Two particles of fluid in homogeneous isotropic turbulence are, at a given instant, a distant r apart. Derive expressions for $\overline{(\frac{dr}{dt})^2}$ and $\overline{\Omega^2}$, where Ω is the angular velocity of the line joining the particles. Hence show that, for values of r small compared with λ,

$$\overline{\left(\frac{dr}{dt}\right)^2} = \frac{r^2u^2}{\lambda^2}$$

and

$$\overline{\Omega^2} = \frac{4u^2}{\lambda^2}.$$

References

[1] Batchelor, G.K., *The Theory of Homogeneous Turbulence*, Cambridge University Press: Cambridge, 1953.

[2] Curle, N. and Davies, H.J., *Modern Fluid Dynamics, Vol. 1: Incompressible Flow*, D. Van Nostrand Company Ltd.: London, 1968.

[3] Lamb, H., *Hydrodynamics*, 6th edn, New York: Cambridge University Press, 1945.

[4] Lighthill, M.J., Fourier Analysis and Generalised functions, Cambridge University Press: Cambridge, 1958.

[5] Longuet-Higgins, M.S., On the nonlinear transfer of energy in the peak of a gravity wave spectrum: a simplified model, *Proc. Roy. Soc. London*, **A347**, pp. 311–328, 1976.

[6] Phillips, O.M., *The Dynamics of the Upper Ocean*, Cambridge University Press: Cambridge, 1966.

[7] Rahman, M., *Water Waves: Relating Modern Theory to Advanced Engineering Applications*, Clarendon Press: Oxford, 1995.

Subject index